Robust Optical Flow Estimation in Outdoor Environments

Robust Optical Flow Estimation in Outdoor Environments

Sana Rao
Bahauddin Zakariya University, Multan, Pakistan

Rao Kashif
National University of Modern Languages, Multan Campus, Pakistan

IEEE Press
WILEY

Published by John Wiley & Sons, Inc., Hoboken, New Jersey.

Library of Congress Cataloging-in-Publication Data applied for

Hardback ISBN: 9781394294220
ePub ISBN: 9781394294237
ePDF ISBN: 9781394294244
oBook ISBN: 9781394294251

Cover Design: Wiley
Cover Image: © lupengyu/Getty Images

Set in 9.5/12.5pt STIXTwoText by Straive, Chennai, India

To the researchers, engineers, and visionaries who inspire innovation – and to my family, who inspire me every day.

Contents

About the Authors

Dr. Sana Rao is a researcher and academic specializing in computer vision, optical flow estimation, and intelligent vision systems operating under challenging outdoor conditions. She received her Ph.D. in Computer Vision, with a research focus on robust motion analysis, human pose estimation, and AI-driven visual computing. Dr. Sana is currently serving as an Assistant Professor in the Department of Information and Communication Technology, Bahauddin Zakariya University, Multan, Pakistan. Her scholarly work bridges theoretical advancements with practical applications, contributing to the development of reliable computer vision systems for robotics, autonomous navigation, and surveillance.

Dr. Rao Kashif is a distinguished academic, researcher, and author with profound expertise in cutting-edge communication technologies. He earned his Ph.D. in Electronic Science and Technology from the prestigious University of Science and Technology of China (USTC), where his research focused on designing high-speed wireless and visible light communication (VLC) hybrid systems. His specialization spans Free Space Optics, Radio over Fiber, VLC, and advanced hybrid communication systems. In addition to his academic pursuits, Dr. Kashif has extensive industry experience, having worked as a Project Control Manager at Huawei Technologies, where he was awarded the prestigious "Future Star Award" for his contributions. He has also served as the Head of the Faculty of Engineering and Computing at the National University of Modern Languages, Pakistan, where he continues to teach and mentor students in advanced engineering and computing disciplines. His work reflects a unique blend of academic excellence, industrial experience, and leadership in higher education.

Preface

The idea for this book grew out of a simple yet challenging question: how can we design intelligent vision systems that continue to perform reliably when faced with unpredictable outdoor conditions? Over the years, we have observed the increasing demand for autonomous vehicles, drones, surveillance systems, agricultural robots, and other smart technologies that depend on accurate motion estimation. At the same time, we noticed a gap in existing literature when it comes to addressing the harsh realities of outdoor environments: bright sunlight, dense fog, heavy rain, snow, or dynamic urban scenes, that often disrupt conventional optical flow methods.

This book, *Robust Optical Flow Estimation in Outdoor Environments*, is our attempt to fill that gap. It brings together theoretical foundations, state-of-the-art techniques, and real-world applications into a single volume. Our aim is to provide readers not only with the concepts but also with practical strategies for building vision-based systems that can operate with confidence in complex environments.

The book is intended for a broad audience: researchers looking for new ideas, engineers working on practical systems, and students seeking a solid foundation in computer vision and robotics. Throughout the chapters, we have tried to balance depth with clarity, integrating illustrative examples, experimental insights, and innovative methodologies. In particular, the book emphasizes the transition from traditional optical flow approaches to modern robust techniques that are designed to cope with outdoor challenges.

What makes this book unique is its focus on real-world relevance. While much of the research in optical flow has been conducted under controlled conditions, our goal has been to highlight how these methods can be adapted and improved for deployment in the field. Whether it is enhancing the safety of autonomous vehicles, improving the efficiency of outdoor surveillance systems, or enabling agricultural robots to operate under variable weather, the applications are both timely and impactful.

We would also like to acknowledge the support, encouragement, and insights we have received along this journey from colleagues, peers in the computer vision and robotics community, and our family, who provided constant motivation. Their contributions, directly and indirectly, have made this work possible.

It is our hope that this book will serve not only as a comprehensive reference but also as a source of inspiration for further research and innovation in robust computer vision systems. If it can help even a small step toward building safer, smarter, and more reliable technologies, we will consider the effort worthwhile.

Acknowledgments

Writing a book is never a solitary effort, and we are deeply grateful to the many individuals and institutions who have supported us throughout this journey.

First and foremost, we wish to thank our colleagues and collaborators in the fields of computer vision and robotics, whose discussions, feedback, and encouragement have enriched the ideas presented in this work. Their expertise and insights have been invaluable in shaping the direction of this book.

We are also indebted to our friends and peers, who provided constant encouragement and motivation during the long process of research and writing. Their patience and support made this effort possible.

Special thanks go to the editorial and production staff at *Wiley*, whose professionalism, guidance, and attention to detail have greatly improved the clarity and presentation of this book. Their commitment has ensured that the work reached its final polished form.

We would also like to acknowledge the inspiration and unwavering support of our family, who stood by me through every stage of this endeavor. Their belief in our work has been our greatest source of strength.

To all those who contributed directly or indirectly to the completion of this book, we extend our heartfelt thanks.

1

Introduction

1.1 Introduction

Optical flow estimation is a fundamental process in computer vision, serving as a cornerstone for numerous applications that rely on the interpretation and analysis of dynamic visual scenes. In essence, optical flow refers to the pattern of apparent motion of objects, surfaces, or edges in a visual field, caused by the relative movement between an observer (e.g., a camera) and the scene being observed. This motion is captured across consecutive frames in a video or image sequence, allowing systems to infer how objects move through space and time. This process is not merely about tracking the visible changes but about comprehensively understanding the motion in the scene. The estimation of optical flow enables machines to perceive motion in a manner that mimics human visual perception, providing vital data for decision-making in real-time systems.

At its core, optical flow determines the motion of each pixel between two consecutive image frames. This pixel-by-pixel analysis allows for the creation of a dense flow field, which illustrates the direction and magnitude of movement within a scene. In each frame of a video, pixels represent specific points in space that may belong to moving objects, stationary backgrounds, or changing lighting conditions. By analyzing the changes in pixel intensity values over time, optical flow algorithms can determine how these pixels have shifted, thereby revealing both the velocity and trajectory of objects. Unlike simple tracking systems that follow individual features, optical flow works on a much finer scale, considering all pixels within the image and providing a continuous motion map. This fine-grained motion data is crucial for complex tasks such as motion segmentation, where different objects or regions moving at various speeds and directions need to be identified separately.

This information is crucial for understanding the behavior of objects, tracking them, and predicting their future movements. For instance, in an autonomous driving system, the accurate estimation of optical flow is essential for predicting

Robust Optical Flow Estimation in Outdoor Environments, First Edition. Sana Rao and Rao Kashif.

the movements of pedestrians, cyclists, and other vehicles. By understanding not only where these objects are but also how they are likely to move in the near future, the system can make decisions to avoid collisions or plan safer routes. Similarly, in the field of medical imaging, optical flow allows for the tracking of internal structures such as organs or blood vessels over time, providing invaluable information for both diagnostics and treatment planning. The ability to predict future movements also plays a critical role in robotics, where machines must anticipate the behavior of dynamic objects in their environment to navigate and interact safely and effectively.

The task of estimating optical flow involves identifying how pixels move across image sequences, which in turn reveals both the speed and direction of moving objects. This process hinges on the assumption that the intensity of pixels remains relatively constant as they move from one frame to the next. By tracking the changes in pixel locations over time, optical flow algorithms generate a velocity field that maps the motion occurring throughout the image. However, achieving high accuracy in this estimation is far from trivial. Factors such as noise, changes in lighting, and distortions caused by camera movement can obscure the true motion of pixels, leading to incorrect or incomplete flow fields. The computational challenge lies in designing algorithms that can robustly handle these variations and extract the underlying motion with precision.

This process can be challenging, particularly in environments where factors like lighting, weather, and occlusion complicate the analysis. In real-world scenarios, the complexity of scenes makes it difficult to maintain the accuracy of optical flow estimations. For example, varying lighting conditions, such as sudden changes in brightness or shadows cast by moving objects, can significantly affect the intensity of pixels. This can result in false motion estimations, where the algorithm detects movement where there is none or fails to recognize actual movement. Weather conditions such as rain, snow, or fog further exacerbate the problem by distorting the clarity of the scene, making it difficult for the algorithm to discern motion accurately. Additionally, occlusions, where one object passes in front of another, pose a significant challenge. When an object is partially or fully obscured, the algorithm must infer its motion based on the surrounding pixels, which can lead to inaccuracies.

For example, accurately estimating motion in scenes with fog, haze, or low-contrast conditions presents significant hurdles. Fog and haze scatter light, reducing the visibility of objects and blurring the edges between moving and stationary elements. In such conditions, the contrast between the moving object and its background diminishes, making it difficult for optical flow algorithms to distinguish one from the other. Low-contrast scenes, which often occur during dawn or dusk, present similar challenges. The lack of clear intensity differences between pixels means that the movement of objects becomes harder to track. As a result, traditional optical flow techniques, which rely on high contrast to detect motion, often struggle in such conditions, leading to significant errors in motion estimation. This has prompted ongoing research into developing more robust methods that can cope with the complexities of low-visibility environments.

Optical flow estimation is employed in various applications that require a deep understanding of how objects move. These applications span a broad range of industries and disciplines, each with its own set of requirements and challenges. For instance, in autonomous vehicles, optical flow is critical for detecting and tracking pedestrians, cyclists, and other vehicles in real time. By understanding the movement of these entities, the vehicle can make decisions about acceleration, braking, and steering to avoid collisions and navigate safely. In the sports industry, optical flow is used to analyze the movements of athletes, providing insights into performance, technique, and potential areas for improvement. In scientific research, optical flow techniques are applied to study fluid dynamics, allowing researchers to visualize and quantify the movement of fluids in experiments and simulations.

Examples include detecting pedestrians in autonomous vehicles, analyzing sports movements, or understanding fluid dynamics in scientific research. In the context of autonomous vehicles, optical flow estimation is crucial for enabling the vehicle to perceive its surroundings and make informed decisions. For example, by analyzing the motion of pedestrians, the vehicle can predict their future movements and determine whether they are likely to cross the road or stay on the sidewalk. This allows the vehicle to adjust its speed or course to avoid potential accidents. Similarly, in sports analytics, optical flow is used to track the movements of athletes, allowing coaches and analysts to break down the biomechanics of an athlete's performance. By understanding the speed, direction, and coordination of movements, optical flow provides valuable data for improving training regimens and enhancing performance. In fluid dynamics research, optical flow is employed to visualize the movement of fluids in experiments, helping scientists better understand the behavior of liquids and gases under various conditions.

However, despite the immense progress in optical flow research, challenges persist, particularly in capturing motion in complex outdoor environments. While significant advancements have been made in developing more robust and accurate optical flow algorithms, there are still numerous limitations that need to be addressed. Outdoor environments, in particular, present a host of difficulties due to the variability in lighting, weather conditions, and occlusions. For instance, in scenarios where there are rapid changes in lighting or where objects are partially obscured, traditional optical flow algorithms often fail to provide accurate motion estimates. Similarly, weather conditions such as rain, snow, and fog can obscure the view of objects, making it difficult to track their movement accurately.

These challenges highlight the need for continued research into more adaptive and resilient optical flow techniques that can cope with the complexities of real-world environments.

1.2 Background and Significance

In the real world, we constantly perceive visual motion in our surrounding environment, a key aspect of how humans understand and interact with the world around them. Whether it is the flow of traffic, the movement of people, or the shifting patterns of light and shadow, motion provides us with critical information about the dynamics of the environment. This visual motion offers rich clues about the behavior, structure, and interaction of objects [1, 2]. For example, a moving vehicle reveals information not just about its location, but also about its speed, direction, and the underlying forces driving its motion. Similarly, the movement of clouds across the sky or the fluttering of leaves in the wind conveys a complex interplay of forces that influence how we perceive and predict future events. Motion gives objects context in both space and time, allowing us to make decisions quickly and efficiently based on their observed behavior.

The motion of objects provides rich information about their dynamic behavior and structure, helping to reveal insights that are critical for both machines and humans to make informed decisions. As humans, we have evolved to interpret motion instinctively: we can anticipate the path of a ball thrown through the air, react to a fast-moving car approaching an intersection, or understand the urgency in someone's stride as they walk toward us. This kind of visual processing is not only fundamental for survival but is also deeply embedded in the way we interact with our surroundings. From a computational perspective, the ability to extract such motion information is equally important for machines to interpret and act upon their environments. This makes motion estimation a key challenge for artificial systems, which must replicate the efficiency and accuracy of human motion perception in tasks such as object detection, tracking, and prediction.

Although we can recognize an object from a static image, motion cues provide more robust and reliable information. For instance, while a stationary car in an image can be identified based on shape and color, its motion over time provides additional information about where it is headed, how fast it is moving, and what obstacles it may encounter. Motion can also reveal hidden attributes of an object that are not evident in still images, such as its momentum, acceleration, or even its intended destination. By tracking the movement of pixels over time, machines can infer a more complete understanding of the scene. This dynamic information can be critical in distinguishing objects that may look similar in still images but behave very differently in motion, such as distinguishing between a person walking versus a person running. Motion cues can provide disambiguation, making optical flow estimation an indispensable tool in many computer vision tasks where recognizing movement patterns is essential.

Therefore, the extraction of motion information or motion estimation from image sequences is a fundamental problem in computer vision, often referred to as optical flow estimation. Optical flow involves the calculation of the motion vectors that describe how each pixel in an image frame moves from one frame to the next. This pixel-by-pixel analysis reveals how objects shift through space, allowing for detailed mapping of the velocity and direction of movement across the entire image. The concept of optical flow has been integral to the development of modern computer vision technologies, serving as the backbone for numerous applications that require an understanding of dynamic scenes. Whether it is enabling a robot to navigate through a crowded environment or helping autonomous vehicles detect pedestrians, optical flow estimation allows machines to analyze and interpret motion with a high degree of precision.

Optical flow estimation is often visualized through flow fields, where vectors depict the movement of pixels between frames. In Figure 1.1, the process of optical flow estimation is illustrated step-by-step: Figure 1.1a, b show the current and reference image frames, which serve as the input for the estimation process. From these frames, an optical flow algorithm calculates the motion of pixels, resulting in the generation of the estimated flow field, as shown in (Figure 1.1c). This flow field is a dense map of motion vectors that describe the displacement of pixels from one frame to the next. Finally, (Figure 1.1d) shows the color-coded ground-truth optical flow, where each pixel's motion is represented by a specific color. In this color-coding system, different hues correspond to different directions of motion, while the saturation of the color represents the speed or magnitude of the flow vector. This visualization makes it easier for humans to intuitively understand the complex patterns of motion in a scene, helping to translate the abstract flow vectors into a more tangible form.

The color-coded optical flow is particularly beneficial for human interpretation, as it translates the motion data into a visual format that is easy to understand. By associating color hues with motion directions and color saturation with motion magnitude, these visualizations create a more intuitive perception of the flow vectors. For example, motion toward the right might be represented by red, while motion upward is blue, and their respective intensities convey speed. Such a visual representation can help both researchers and practitioners to quickly identify patterns, outliers, or anomalies in motion data. This is particularly important when working with large datasets or complex environments where manually interpreting motion vectors might be difficult. The use of color coding enables an immediate visual grasp of the motion flow, making it easier to detect inconsistencies,

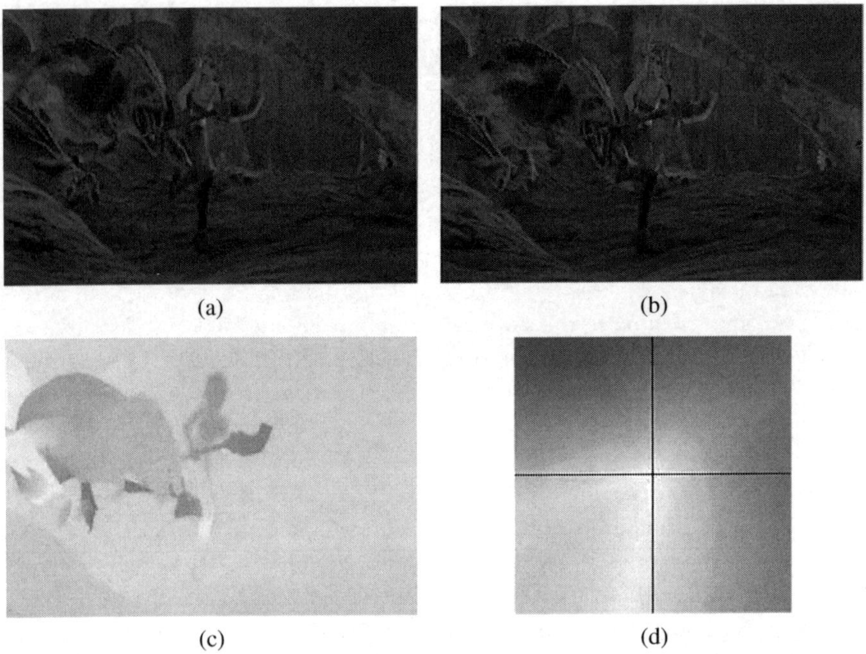

Figure 1.1 The MPI-Sintel image sequence and its ground-truth flow with color-code for the flow field. Left to right: (a) the first image frame, (b) the second image frame. *Source:* Max Planck Institute for Intelligent Systems / http://sintel.is.tue.mpg.de/ / last accessed on Nov 25, 2025, (c) the ground-truth of the optical flow, and (d) the color-code for the optical flow field. *Source:* Max Planck Institute for Intelligent Systems / http://sintel.is.tue.mpg.de/ / last accessed on October 25, 2025.

such as areas where motion is unexpectedly slow or fast, or where objects may be behaving abnormally.

Optical flow estimation plays a crucial role in various computer vision tasks, including motion segmentation, object tracking, and structure-from-motion. In motion segmentation, optical flow is used to partition an image sequence into distinct regions based on their motion patterns. For instance, it can separate moving objects, such as vehicles or pedestrians, from static backgrounds like buildings or trees. This is particularly useful in autonomous driving, where distinguishing between dynamic and static elements is vital for safe navigation. In object tracking, optical flow provides continuous monitoring of an object's position, allowing systems to follow its movement frame by frame. This is critical for applications like surveillance, where tracking the motion of individuals or vehicles over time is required. Finally, in structure-from-motion tasks, optical flow helps in reconstructing the three-dimensional structure of a scene from two-dimensional images, enabling machines to understand depth and spatial relationships between objects.

However, despite the remarkable advances in optical flow estimation, there remain several unresolved challenges, particularly in outdoor environments where factors like haze, fog, and occlusion complicate motion perception. In outdoor scenes, optical flow estimation can be significantly affected by environmental factors that distort the clarity and visibility of objects. For instance, in hazy or foggy conditions, the diffusion of light scatters the pixels across an image, making it difficult to detect clear edges or define the boundaries of moving objects. These visual disruptions can cause motion vectors to be miscalculated, leading to inaccuracies in flow estimation. Similarly, occlusion, where one object passes in front of another, poses an additional challenge, as the algorithm must predict the movement of the obscured object based on incomplete data. This has driven ongoing research into more robust optical flow techniques that can handle such challenges, particularly for real-world applications like autonomous navigation, surveillance, and environmental monitoring.

Given a sequence of images captured in outdoor scenes with multiple moving objects, the goal is to estimate the motion of each pixel between two consecutive frames and segment the scene into the foreground (containing moving objects) and background (the static scene). The separation of foreground and background is critical in many computer vision tasks, such as object detection, activity recognition, and scene understanding. For example, in autonomous driving, distinguishing between moving objects (such as vehicles and pedestrians) and stationary elements (such as road signs and buildings) enables the system to focus its attention on relevant dynamic elements. Optical flow is at the heart of this task, allowing for a pixel-level understanding of motion and scene structure. This segmentation process not only aids in identifying objects but also provides context for how these objects interact with one another within the scene.

1.3 Optical Flow Estimation Applications

Optical flow calculation estimates the motion information of objects between neighboring frames, and its basic assumption is to provide information about the direction as well as the velocity of every pixel, which is important for a variety of applications in computer vision. Next, we will discuss several applications of motion estimation, which require robust optical flow estimation.

1) **Autonomous Vehicles:** In autonomous vehicles, optical flow plays a pivotal role in detecting obstacles, predicting future movements, and supporting

real-time decision-making. Self-driving cars rely heavily on accurate motion information to track the movements of other vehicles, pedestrians, cyclists, and even road surface changes. By estimating the velocity and direction of these elements, optical flow provides essential input for advanced driving algorithms, which use this data to plan safe and efficient routes. Whether it is anticipating a pedestrian crossing the road or calculating the speed of an oncoming car, optical flow estimation helps autonomous systems ensure smooth and secure navigation. In this context, even minor errors in motion estimation can result in significant risks, making optical flow accuracy a cornerstone of autonomous vehicle safety and performance.

2) **Augmented and Virtual Reality (AR/VR):** In AR and VR systems, precise real-time motion tracking is crucial for maintaining an immersive user experience. Optical flow allows these systems to capture and process user movements and environmental changes fluidly, ensuring that virtual objects move smoothly and naturally. In augmented reality, virtual elements must seamlessly blend with the real-world environment, and any lag or inaccuracies in motion tracking can break the user's sense of immersion. For virtual reality gaming, where users interact with fully digital environments, even small delays in motion estimation can disrupt gameplay and cause discomfort. By leveraging optical flow estimation, AR and VR applications can synchronize virtual objects with real-world or digital movements, ensuring that user interactions are as lifelike as possible.
3) **Medical Imaging:** In medical imaging, optical flow is applied to track the motion of tissues and organs, offering invaluable insights during surgical procedures. For example, in cardiac surgery, the precise real-time tracking of the heart's motion is critical for the surgeon's ability to navigate around delicate tissues and make informed decisions during the operation. Optical flow estimation enables systems to capture the motion of specific pixels corresponding to different parts of the organ, allowing for highly targeted interventions. This motion data is also useful in other medical contexts, such as tracking lung movement in respiratory analysis or the motion of tissues in noninvasive diagnostic imaging. By enhancing the precision of motion tracking in medical applications, optical flow supports more accurate diagnoses and improved patient outcomes.
4) **Visual Surveillance:** In visual surveillance systems, optical flow estimation is crucial for detecting and tracking moving objects, such as vehicles or people [2–5]. This technology works by identifying differences in motion between the current frame and a reference frame at the pixel level. This allows the system to effectively segment moving objects from their background and analyze their behavior. For example, a surveillance system might use optical flow to detect suspicious activity in a crowded area or to monitor traffic patterns in

real time. Because optical flow can detect motion at a very fine-grained level, it is well-suited for motion-based object detection, which is a core feature of many modern surveillance systems. These systems depend on the accuracy of optical flow to achieve reliable object detection and event monitoring, especially in crowded or complex environments.

5) **Robot Navigation Systems:** Optical flow is also vital for robot navigation, enabling robots to avoid obstacles and navigate autonomously from a start point to a goal point. In robotics, there are generally two categories of navigation: known environments, where the robot has prior knowledge of its surroundings, and unknown environments, where it must rely on sensor data to make navigation decisions. Optical flow plays a crucial role in both contexts by helping robots detect and avoid obstacles in their path. Inspired by the natural navigation abilities of birds and bees, which use visual cues to fly through narrow spaces without colliding, robots utilize optical flow to gather environmental data and determine their speed and direction. This motion information allows robots to adjust their trajectory dynamically, avoiding collisions and navigating efficiently [6–8]. The adaptability and precision provided by optical flow make it an invaluable tool in the development of autonomous robotics [9, 10].
6) **Action Recognition and Video Classification:** Optical flow estimation is one of the most effective techniques for action recognition and video classification because it captures the dynamic behavior of objects over time. Motion is one of the most critical low-level cues for recognizing actions, such as walking, running, or jumping, in video sequences. Sparse motion points alone can provide enough information to identify these actions. Studies, such as the work in [11], demonstrate that using optical flow for action recognition significantly improves classification accuracy compared to raw image input alone. Optical flow is particularly valuable because it remains robust against appearance changes, such as variations in lighting, pose, or background, which can disrupt other motion detection techniques [12–14]. These findings underscore how accurate optical flow estimation enhances the performance of high-level vision tasks like action recognition and video analysis.

In summary, the accuracy of optical flow estimation is a critical factor in all the aforementioned applications. Whether it is navigating a self-driving car through busy streets or ensuring the precise tracking of organ movement in surgery, the ability to accurately estimate motion enhances both the reliability and robustness of these systems. As a result, the development of more precise and efficient optical flow algorithms continues to be a primary focus of research in computer vision. The more accurate the optical flow estimation, the more reliable the segmentation, tracking, or analytical tasks in which it is applied.

1.4 Current Research Trends and Problems

Optical flow has emerged as a powerful tool in various fields, enabling advancements in motion estimation and scene understanding. However, as research in this area progresses, a series of inherent challenges and limitations continue to be faced by researchers. These difficulties primarily arise from the foundational assumptions made in optical flow estimation methods, particularly when applied to complex real-world scenarios.

Brightness Constancy: Most optical flow estimation methods are based on the brightness constancy assumption, which is an ill-posed problem. The brightness constancy assumes that the brightness of moving objects remains constant. This is a reasonable assumption when the illumination conditions are constant and only a small amount of motion between image frames exists. The violation of the brightness constancy assumption leads to an aperture problem that denotes the ambiguity in motion perception.

Example: Imagine a video sequence recorded during sunset, where the illumination changes rapidly from frame to frame. The brightness constancy assumption would struggle in this scenario, leading to inaccurate motion estimation due to the constantly changing lighting conditions. This issue is compounded when there is significant motion between frames, making it difficult to differentiate actual motion from changes in brightness.

The failure of the brightness constancy assumption often results in what is known as the aperture problem a phenomenon where the perceived motion direction of objects is ambiguous, especially when viewed through small "apertures" or observation windows.

Motion Boundaries and the Smoothness Assumption: A typical way to overcome the above ill-posed problem is to utilize an additional smoothness assumption. Neighboring pixels usually belong to the same surface in the three-dimensional world, and they should have similar motions. Motion boundaries may violate the smoothness assumption because that the neighboring pixels belonging to different surfaces may have different motions. The smoothness condition in these areas blurs the motion boundaries, see Figure 1.2.

Example: In a video of a moving car on a highway, the pixels representing the car and the background should have different motions. However, conventional methods may blur the boundary between the car and the road, leading to poor motion segmentation.

Optical Flow Estimation for Low-contrast Image Sequences: The real-world image sequences may contain many challenges, such as motion blur,

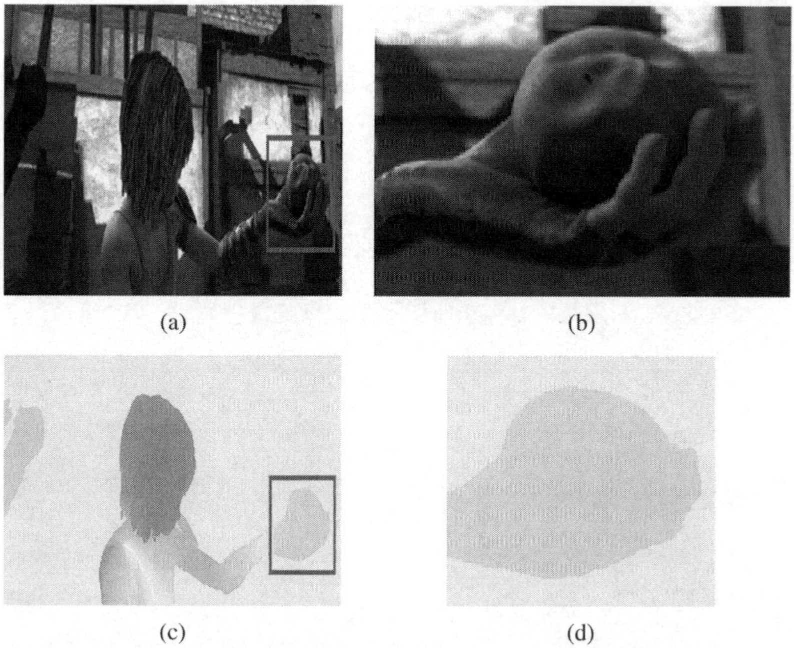

Figure 1.2 Smoothness across the motion boundaries destroys the motion structure. Left to right: (a) the input image frame. *Source:* Max Planck Institute for Intelligent Systems / http://sintel.is.tue.mpg.de/ / last accessed on Nov 25, 2025, (b) an amplified view of the marked area, (c) the optical flow estimation, and (d) an amplified view of the flow field. *Source:* Sana Rao (2024) / Springer Nature.

illumination changes, and bad weather conditions. Currently, some optical flow estimation methods only consider a few conditions. Most state-of-the-art motion estimation methods cannot effectively handle image sequences captured under complex outdoor scenarios well, see Figure 1.3.

The low-contrast images may cause the violation of the brightness constancy assumption. It is important to propose a robust method that can handle low-contrast sequences to obtain accurate optical flow estimation.

1.4.1 Key Challenges and Solutions

Variational methods play a very important role in optical flow estimation. Some modern methods are developed based on variational methods and achieve promising performance.

Recently, the total variation (TV)-L^1-based methods [15] have become more popular in motion estimation due to their good performance in preserving the

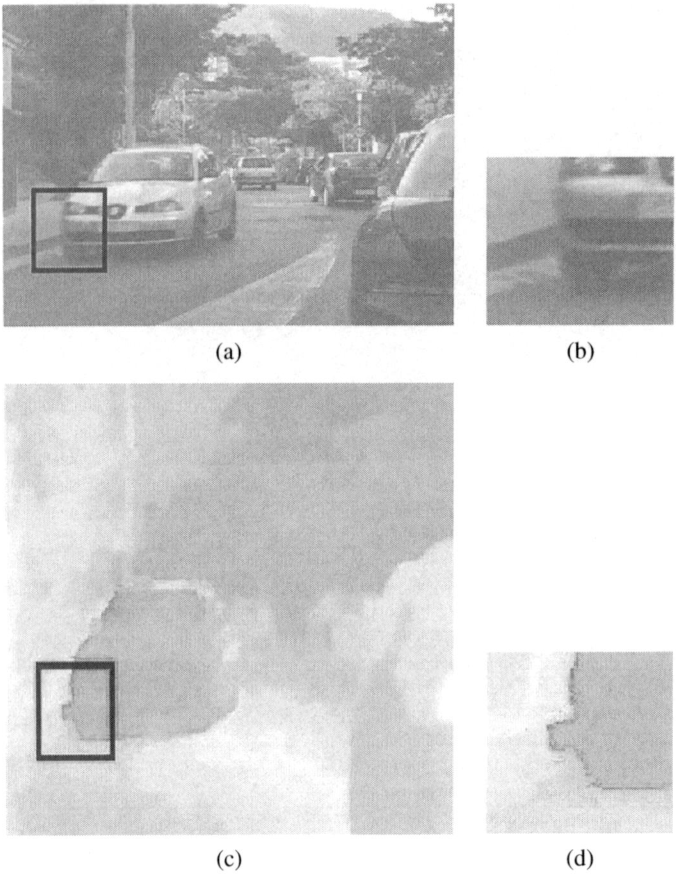

Figure 1.3 Low-contrast images limit the visibility and make the optical flow estimation task challenging. Left to right: (a) the input image frame, (b) an amplified view of the marked area, (c) the optical flow estimation, and (d) an amplified view of the flow field. *Source:* Rao Kashif et al., (2022) / with permission of IEEE.

boundaries. A structure texture decomposition with TV is proposed to enhance the robustness by applying the L^1 norm in the data term in [16].

A non-local term is introduced in [17] with the L^1 norm to overcome the effects of poorly textured regions and small structures on TV-L^1. The intuition is that if we know other pixels belong to the same surface, we can weight them accordingly and thus alleviate the problem. It is observed that the weighted median filter relies more on the divergence of the motion flow rather than just the brightness distance, and a robust optical flow estimation method is proposed in [18] for computing the weight of pixels in a filtering window.

Most optical flow-based segmentation methods are sensitive to illumination changes due to the brightness constancy assumption, or they may suffer from the boundary blurring problem [19]. Recently, layer-based methods [20] have shown their robustness for motion segmentation, where the flow field is assumed to be already estimated and directly used for segmentation. The combination of motion segmentation and alpha matting within a graph cut framework is proposed in [21] to achieve robust motion layer segmentation. The low-contrast problem is tackled in [22] for moving object segmentation based on background modeling, and the performance depends on the dynamic matrix.

1.5 Structure of the Book

This book is organized into eight chapters, each delving into different aspects of optical flow estimation and its applications in various domains. Below is a brief overview of each chapter.

Chapter 2: This chapter provides an in-depth overview of classical and contemporary methods for optical flow estimation and motion segmentation. The discussion begins with the variational optical flow estimation method introduced by Horn and Schunck [23], which forms the basis for many modern approaches. We also explore the TV optical flow estimation method [24], widely adopted due to its ability to preserve motion boundaries.

The chapter discusses the limitations of current methods in real-world conditions and presents the evaluation measures and datasets used in optical flow research. This foundational knowledge sets the stage for the more advanced techniques introduced in subsequent chapters.

Chapter 3: In this chapter, we introduce a robust optical flow estimation method designed for complex outdoor scenarios. The proposed method, based on nonlocal total variation (NLTV-L^1), employs a new support-weight function to preserve edges and motion boundaries effectively.

Additionally, we apply the estimated flow field to the task of motion segmentation, presenting a motion pattern-based background–foreground layer segmentation technique to accurately distinguish moving objects from their surroundings in dynamic environments.

Chapter 4: This chapter proposes an optical flow estimation technique that incorporates a weighted guided filter to handle ill-defined edges and boundaries. By integrating this filter into the NLTV-L^1 method, we enhance the preservation of motion boundaries, which is critical for accurate motion estimation.

Experimental evaluations demonstrate the efficacy of this method, especially for scenarios requiring edge preservation. We also explore a unique application by using the method to study the motion of ants, a complex task due to the intricate social behaviors of these creatures. The optical flow-based analysis outperforms manual tracking methods.

Chapter 5: In this chapter, we present an advanced optical flow estimation method using a non-local steering kernel. This method is designed to minimize the impact of outliers while maintaining motion boundary integrity, making it more robust for complex motion estimation tasks.

The proposed method is evaluated on benchmark datasets, and comparisons with other state-of-the-art techniques demonstrate its superior performance in preserving motion boundaries and managing complex motion scenarios.

Chapter 6: This chapter explores the latest deep learning-based approaches for optical flow estimation. We review various convolutional neural network (CNN) architectures and transformer models designed for optical flow tasks. These models offer significant improvements in both accuracy and computational efficiency.

Furthermore, we discuss the challenges and solutions in applying deep learning to optical flow, including the use of synthetic datasets for training and the transfer of learned representations to real-world scenarios. The chapter provides a comprehensive analysis of the strengths and limitations of deep learning approaches in optical flow estimation.

Chapter 7: This chapter focuses on the application of optical flow estimation in real-time systems and autonomous vehicles. We discuss the computational challenges of integrating optical flow algorithms into systems that require real-time processing and present optimized techniques that meet these constraints.

The chapter explores the role of optical flow in perception systems for autonomous vehicles, particularly in tasks such as obstacle detection, motion prediction, and navigation. Case studies and real-world implementations highlight the importance of robust and efficient optical flow methods for ensuring the safety and reliability of autonomous systems.

Chapter 8: In the final chapter, we summarize the key contributions made throughout the book, highlighting the advancements in optical flow estimation techniques and their applications in various fields.

We also outline potential future research directions, such as further optimizing deep learning models for real-time applications, improving robustness in extreme conditions, and exploring new datasets for training and evaluating optical flow methods. The chapter concludes by discussing the broader implications of optical flow research for autonomous systems, computer vision, and artificial intelligence.

Examples with Solutions

Brightness Constancy Assumption in Practice

Problems:
Suppose a white ball moves across a dark background from left to right in consecutive video frames. The ball's brightness intensity value is 220 (on a scale of 0–255), and the background is 30. How can we use the brightness constancy assumption to estimate motion?

Solution:
In frame 1, the ball is at pixel location $(x, y) = (50, 100)$.

In frame 2, the same brightness value (220) appears at $(x + 5, y)$.

By comparing the pixel intensity across frames, we determine that the displacement is $\Delta x = 5$ pixels, $\Delta y = 0$.

The velocity vector for this object = (5, 0).

Interpretation:
This simple calculation shows how pixel intensity tracking allows an algorithm to detect motion, assuming brightness is constant. However, if the lighting changed (say brightness drops to 180), the assumption breaks, leading to errors, motivating more advanced techniques.

Optical Flow in Autonomous Driving

Problems:
An autonomous car captures two consecutive frames of a road scene. A pedestrian moves 3 pixels to the right and 1 pixel up between frames, while a tree in the background does not move. How does optical flow separate foreground (pedestrian) from background (tree)?

Solution:
The car's camera records pixel positions:

- **Pedestrian:** Frame 1 → (120, 200); Frame 2 → (123, 199).
- **Tree:** Frame 1 → (300, 400); Frame 2 → (300, 400).

Pedestrian displacement = (3, −1).
Tree displacement = (0, 0).
Foreground–background segmentation:

- Moving object (pedestrian) detected.
- Static object (tree) classified as background.

Interpretation:
This example shows how optical flow distinguishes dynamic versus static objects, which is critical for autonomous navigation.

Motion Boundary Problem

Problem:
Consider a video where a red car moves at velocity (4, 0) pixels/frame, and the road beneath it is static. What happens at the boundary between the car and the road when using the smoothness assumption?

Solution:

- Smoothness assumption expects neighboring pixels to have similar motion.
- Car pixels $\rightarrow (4, 0)$, Road pixels $\rightarrow (0, 0)$
- At the boundary, the algorithm blends $\rightarrow$ estimated motion $\approx (2, 0)$.
- This blurring causes errors: part of the road may appear as if it is moving!

Interpretation:
This demonstrates the motion boundary challenge, which motivates advanced methods like TV regularization to preserve sharp object boundaries.

Low-contrast Challenge

Problem:
A drone is flying at dawn. The brightness values of the background sky = 100, while a flying bird has brightness = 110. Between two frames, the bird shifts by 5 pixels. Can the optical flow algorithm detect the motion?

Solution:

- Brightness difference between object and background = only 10 units.
- In low-contrast conditions, noise (sensor error ± 8 units) overlaps with motion signal.
- The motion vector (5 pixels) is weak and may be lost in noise.
- Advanced filters (like guided filters) or deep learning models are needed.

Interpretation:
This shows why **low contrast = low detectability**, motivating robust outdoor flow estimation.

Visualizing Optical Flow

Problem:
We want to visualize motion between two frames of a football match. A player runs to the right at velocity (6, 0). How is this represented in a color-coded optical flow field?

Solution:

- Direction (right) → mapped to red hue.
- Magnitude (6 pixels/frame) → mapped to brightness level.
- The faster the motion, the more saturated the red color appears.
- On the flow map, stationary objects (goalpost, field) remain black/neutral.

Interpretation:
This shows how color coding makes motion interpretable for both humans and machines.

Sports Analytics

Problem:
In a basketball game, a player runs across the court in 12 frames, moving from $x = 50$ to $x = 170$ pixels. What is his average velocity?

Solution:

- Displacement = 170–50 = 120 pixels.
- Frames = 12 → assume 30 fps = 12 ÷ 30 = 0.4 seconds.
- Velocity = 120 ÷ 0.4 = 300 pixels/sec.
- If scaling factor = 0.02 m/pixel → player speed = $300 \times 0.02 = 6$ m/s.

Interpretation:
Optical flow converts raw pixel shifts into real-world performance metrics.

References

1 Tu, Z. (2015). Variational optical flow algorithms for motion estimation. Doctoral dissertation, University Utrecht.
2 Wang, S., Guo, Q., Xu, S., and Su, D. (2021). A moving target detection and localization strategy based on optical flow and pin-hole imaging methods using

monocular vision. *IEEE International Conference on Real-time Computing and Robotics*, 147–152. 15–19 July 2021, Xining, China.

3 Xiao, Y., Gao, G., Wang, L., and Lai, H. (2022). Optical flow-aware-based multi-modal fusion network for violence detection. *Entropy* 24 (7): 939–955.

4 Wu, M., Li, X., Liu, C. et al. (2019). Robust global motion estimation for video security based on improved k-means clustering. *Journal of Ambient Intelligence and Humanized Computing* 10 (2): 439–448.

5 Al-Battal, A.F., Lerman, I.R., and Nguyen, T.Q. (2022). Object detection and tracking in ultrasound scans using an optical flow and semantic segmentation framework based on convolutional neural networks. *IEEE International Conference on Acoustics, Speech and Signal Processing*, 1096–1100. May 22–27, 2022, Singapore, Sans Expo & Convention Centre.

6 Gul, F., Rahiman, W., and Nazli, A.S.S. (2019). A comprehensive study for robot navigation techniques. *Cogent Engineering* 6 (1): 1632046–1632072.

7 Serres, J.R. and Ruffier, F. (2017). Optic flow-based collision-free strategies: from insects to robots. *Arthropod Structure & Development* 46 (5): 703–717.

8 Diamantas, S. and Dasgupta, P. (2022). Optical flow-based place recognition: bridging the gap between simulation and real-world experiment. *IEEE Annual Computing and Communication Workshop and Conference*, 0619–0624. 26–29 January, 2022, Las Vegas, USA.

9 Cho, G., Kim, J., and Oh, H. (2019). Vision-based obstacle avoidance strategies for mavs using optical flows in 3-D textured environments. *Sensors* 19 (11): 2523–2539.

10 Cheng, J., Sun, Y., Chi, W. et al. (2018). An accurate localization scheme for mobile robots using optical flow in dynamic environments. *IEEE International Conference on Robotics and Biomimetics*, 723–728. 12–15 DEC, 2018. Kuala Lumpur Convention Centre, Kuala Lumpur, Malaysia.

11 Tang, Y., Ma, L., and Zhou, L. (2019). Hallucinating optical flow features for video classification, arXiv preprint: 1905.11799.

12 Piergiovanni, A. and Ryoo, M.S. (2019). Representation flow for action recognition. *IEEE Computer Society Conference on Computer Vision and Pattern Recognition,* 9945–9953. 15–20 June, 2019, Long Beach, CA, USA.

13 Chi, L., Tian, G., Mu, Y., and Tian, Q. (2019). Two-stream video classification with cross-modality attention. *IEEE International Conference on Computer Vision Workshops,* 1–6. 27–28 October 2019, Seoul, Korea.

14 Sevilla-Lara, L., Liao, Y., Güney, F. et al. (2018). On the integration of optical flow and action recognition. *German Conference on Pattern Recognition*, 281–297. 10–12 October 2018 Stuttgart.

15 Garamendi, J.F., Lazcano, V., and Ballester, C. (2019). Joint tv-l1 optical flow and occlusion estimation. *Image Process. Line* 9 (1): 432–452.

16 Zach, C., Pock, T., and Bischof, H. (2007). A duality based approach for real-time tv-l1 optical flow. *Joint Pattern Recognition Symposium*, 214–223. 12–14 September, Heidelberg, Germany.

17 Sun, D., Roth, S., and Black, M.J. (2010). Secrets of optical flow estimation and their principles. *IEEE Computer Society Conference on Computer Vision and Pattern Recognition*, 2432–2439. 13–18 June, 2010, San Francisco, CA.

18 Zhang, C., Chen, Z., Wang, M. et al. (2017). Robust non-local tv-l1 optical flow estimation with occlusion detection. *IEEE Transactions on Image Processing* 26 (8): 4055–4067.

19 Anthwal, S. and Ganotra, D. (2019). An overview of optical flow-based approaches for motion segmentation. *The Imaging Science Journal* 67 (5): 284–294.

20 Sevilla-Lara, L., Sun, D., Jampani, V., and Black, M.J. (2016). Optical flow with semantic segmentation and localized layers. *IEEE Computer Society Conference on Computer Vision and Pattern Recognition,* 3889–3898. 27–30 June 2016, Las Vegas, NV, USA.

21 Xiao, J. (2013). Motion segmentation and matting by graph cut. In: *Graph-based Methods in Computer Vision: Developments and Applications* (ed. X. Bai et al.), 95–117. IGI Global Scientific Publishing.

22 Soumya, T. (2008). A moving object segmentation method for low Illumination night videos. *World Congress on Engineering and Computer Science,* 1161–1165. 22–24 October 2008, San Francisco.

23 Horn, B.K. and Schunck, B.G. (1981). Determining optical flow. *Artificial Intelligence* 17 (1–3): 185–203.

24 Sánchez, P.J., Meinhardt-Llopis, E., and Facciolo, G. (2013). Tv-l1 optical flow estimation. *Image Process Line* 3 (1): 137–150.

2

Fundamentals of Optical Flow

2.1 Classical Optical Flow Estimation Models

2.1.1 The Aperture Problem

The aperture problem is a fundamental challenge in optical flow estimation, particularly when the motion is observed through a restricted local window or "aperture." This issue arises because motion perpendicular to the gradient cannot be determined accurately within small regions, especially for objects with repetitive textures, such as horizontal lines moving horizontally. Horizontal motion is difficult to detect within a narrow aperture, which can lead to ambiguous or incorrect motion estimates. This problem becomes pronounced in scenes with uniform or highly structured surfaces where local gradients provide insufficient information for reliable motion estimation.

To address this, researchers introduced global smoothness constraints, which assume that the motion of pixels within a certain neighborhood is similar. By considering the motion of neighboring pixels, the global constraint helps in resolving the ambiguity that arises from local observations alone. This smoothness assumption, along with advanced regularization techniques, enables more accurate motion estimation across entire scenes, even when local details are ambiguous or obscured.

In practical terms, the aperture problem underscores the need for combining both local motion cues and global scene understanding in optical flow algorithms to improve their robustness across a wide range of scenarios, from structured environments to more complex dynamic scenes.

2.1.2 Brightness Constancy Assumption

The basic assumption of optical flow estimation is the brightness constancy assumption, which assumes that the brightness of a moving pixel remains

Robust Optical Flow Estimation in Outdoor Environments, First Edition. Sana Rao and Rao Kashif.

constant over time [1–3]. Mathematically, the brightness assumption can be formulated as:

$$I_1(x, y, t) = I_2(x + u, y + v, t + 1), \tag{2.1}$$

where I_1 and I_2 are the current and the reference frames of the input image sequence in time t and $t+1$, (x, y) denote the horizontal and vertical pixel indices, and (u, v) represents the horizontal and vertical motion for pixel (x, y), respectively.

This assumption, while forming the backbone of many optical flow algorithms, is often violated in real-world scenarios. Environmental factors such as changing illumination, shadows, or reflections can lead to significant deviations in pixel intensity between consecutive frames. For example, outdoor scenes subject to changing light conditions may exhibit brightness variations that cannot be accounted for by the brightness constancy assumption alone. This introduces challenges in accurately estimating motion in such environments.

In addition to illumination changes, the aperture problem discussed earlier compounds the difficulties in applying the brightness constancy assumption [4–7]. Since the flow vectors (u, v) represent the motion of a pixel, and there are two unknowns but only one equation (as shown in Eq. (2.1)), the motion can only be determined in the gradient direction, leaving ambiguities in the orthogonal direction. Addressing these limitations requires augmenting the basic assumptions with additional constraints, as discussed in subsequent sections.

2.2 Variational Optical Flow Estimation Methods

The variational approach to optical flow estimation, introduced by Horn and Schunck, formulates the problem as one of energy minimization. These methods aim to compute an optical flow field that minimizes a global energy function. The energy function is typically composed of two key terms: the data term and the regularization (or smoothness) term.

The data term enforces the brightness constancy assumption, ensuring that the observed motion in the image adheres to the pixel intensities across frames.

The Horn and Schunck model is foundational, but it is susceptible to several challenges when applied to real-world video sequences. For instance, motion boundaries (where objects with different motions intersect) are not well preserved due to the global smoothness constraint, and the brightness constancy assumption is often violated in environments with dynamic lighting conditions. These limitations highlight the need for more adaptive regularization techniques and data terms that can handle real-world complexities such as varying brightness and nonrigid object motion.

2.2.1 Smoothness Constancy Assumption

An additional assumption is needed to incorporate with brightness constancy assumption to make the problem well-posed [8–10]. We need to pool the information from spatial neighbors (i.e., the prior knowledge about the displacement of the neighboring pixels) to overcome the ill-posed problem. Global methods usually introduce a smoothness assumption to solve the aperture problem. The smoothness assumption is based on the condition that neighboring pixels come from the same surface and they tend to have the same type of motion [11].

The smoothness assumption can be written as:

$$u_x^2 + u_y^2 + v_x^2 + v_y^2 = 0. \tag{2.2}$$

Horn and Schunck combine a brightness constancy assumption (i.e., data term) with a smoothness assumption to formulate an energy function to compute optical flow:

$$E(u,v) = \int_{\Omega} \left\{ \underbrace{\left(I_x u + I_y v + I_t\right)^2}_{\text{Data term}} + \lambda \underbrace{\left(u_x^2 + u_y^2 + v_x^2 + v_y^2\right)}_{\text{Smoothness term}} \right\} dxdy, \tag{2.3}$$

where λ is a smoothness parameter to control the smoothness between the two terms. However, these assumptions may be violated across the motion boundaries and for illumination changes. Real-world image sequences contain various challenges [12–15], and a fixed smoothness parameter is unsuitable for controlling the balance between the data and the smoothness terms. For example, at motion boundaries, where objects with different velocities meet, enforcing smooth motion leads to over-smoothing, causing the motion boundary to blur. Conversely, in smooth regions with gradual motion, a stronger smoothness constraint is beneficial. Therefore, modern methods seek to adaptively adjust the smoothness parameter based on local image content to better preserve motion discontinuities while maintaining coherence in homogeneous areas.

In real-world applications, optical flow estimation faces numerous challenges. Sequences captured in dynamic environments, such as outdoor scenes with variable lighting conditions, often violate both the brightness constancy and smoothness assumptions. Variations in illumination, shadows, reflections, and occlusions present significant obstacles for traditional optical flow algorithms. Additionally, fast-moving objects or objects with complex nonrigid motion further complicate the estimation process.

To tackle these challenges, researchers have proposed various robust extensions to the classical models. These include adaptive regularization strategies, multi-scale approaches, and nonlocal methods that leverage information from distant parts of the image to refine the flow estimates. Such advances aim to make optical flow estimation more robust to real-world variations, enabling its use in applications like autonomous driving, video surveillance, and augmented reality (AR).

2.3 Total Variation (TV-L^1) Optical Flow Estimation

The total variation (TV-L^1) optical flow estimation method stands out as an advanced approach for handling motion estimation tasks in complex environments, where traditional methods like the Horn–Schunck algorithm often struggle. The robustness of TV-L^1 stems from its unique ability to preserve edges and sharp discontinuities in the optical flow field. This capability is vital in real-world scenarios where motion boundaries are common, and accurate flow estimation can be compromised by environmental factors such as noise and texture changes. By employing the L^1 norm instead of the L^2 norm, TV-L^1 offers a more resilient framework, minimizing the amplification of errors commonly associated with the brightness constancy assumption in L^2-based methods. Moreover, TV-L^1 methods leverage the principle of structure-texture decomposition to enhance optical flow robustness in the presence of illumination changes, as often encountered in adverse outdoor conditions. This method's ability to strike a balance between motion estimation and edge preservation makes it a crucial tool in modern optical flow research and applications.

2.3.1 Comparison Between TV-L^1 and L^2 Norms

Sun et al. compare several variational optical flow estimation methods and discover that the median filter is a key to improve the accuracy of the variational optical flow estimation methods. In addition, the weighted median filtering method is introduced in the nonlocal term, and the objective function is minimized using the coarse-to-fine strategy. However, it is difficult to assign a reasonable weight to pixels in adverse noise. Li et al. [16] propose a method that can handle the optical flow estimation problem in the rainy scene and utilize a residue channel as a data term to reduce the effect of rain streaks. They also propose a decomposition method to overcome the challenging problem of rain accumulation.

Although the variational optical flow estimation methods can increase the accuracy, robust motion estimation under outdoor complex scenes is still challenging. Several methods devote to handling the abovementioned problems and allow discontinuities in the flow field. Among these methods, TV-L^1 [41–44] shows robustness in reducing the influence of noise. It overcomes the limitation of the variational optical flow estimation methods by changing the quadratic factors, as shown in Eq. (2.3).

The main idea can be formulated as the following energy function:

$$E(u, v) = \int_{\Omega} \{|I_x u + I_y v + I_t| + \lambda(|\nabla u| + |\nabla v|)\} dxdy, \tag{2.4}$$

where |.| denotes the L^1 norm. In Eq. (2.3), since the L^2 norm squares and magnify the error of constancy assumption, the L^1 norm in Eq. (2.4) is more suitable to minimize this error by changing the quadratic factors.

A structure-texture decomposition method with the TV [17–19] is proposed to enhance the robustness of the optical flow estimation. However, such a way is not very effective since the global illumination in adverse outdoor scenarios affects the texture and the structure of an image. Although the TV-L^1-based methods are beneficial for motion estimation in terms of edge preservation and noise removal, they still suffer from the problem of over-smoothing the uninteresting structures in images. The bilateral filtering method has been used in combination with the TV-L^1 [20] to overcome the abovementioned problems.

2.3.2 Non-TV-L^1 Optical Flow Estimation Methods

In recent years, nonlocal total variation (NLTV) methods have been introduced to capture long-range dependencies within the image, allowing for better performance in highly textured or repetitive areas. These methods use the idea of self-similarity, where pixels with similar intensities are more likely to have similar flow vectors, even if they are distant from each other in the image.

A nonlocal term is introduced in [21, 22] with the L^1 norm to overcome the effects of poorly textured regions and small structures on the TV-L^1 method. The nonlocal term is defined as below:

$$E_{NL}(u, v) = \iint_{N_{(x,y)}} \left(\left| u_{x,y} - u_{x',y'} \right| + \left| v_{x,y} - v_{x',y'} \right| \right) dx'dy', \tag{2.5}$$

where $N_{(x,y)}$ denotes the set of neighboring pixels of the center pixel (x, y). In the nonlocal term, $(u_{x,y}, v_{x,y})^T$ is the flow of the center pixel, and $\left(u_{x',y'}, v_{x',y'}\right)^T$ denotes the flow of any neighboring pixel of the center pixel. Usually, it is challenging to optimize Eq. (2.5) directly. Therefore, an auxiliary flow field is usually used to relax the objective function as:

$$E_{NL}(u, v) = \iint_{N_{(x,y)}} \left(\left| \hat{u}_{x,y} - \hat{u}_{x',y'} \right| + \left| \hat{v}_{x,y} - \hat{v}_{x',y'} \right| \right) dx'dy', \tag{2.6}$$

where $(\hat{u}, \hat{v})$ denotes an auxiliary flow field. However, it is observed that the abovementioned solution may result in the over-smoothing problem. Therefore, in [23], the authors introduce weight to the nonlocal term. The intuition is that if we know other pixels belong to the same surface, we can weight them accordingly by which we can alleviate the problem. The improved nonlocal term $E_{NL}(u, v)$ is written as follows:

$$E_{NL}(u,v) = \iint_{N_{(x,y)}} w_{x,y}^{x',y'} \left(| \hat{u}_{x,y} - \hat{u}_{x',y'} | + \left| \hat{v}_{x,y} - \hat{v}_{x',y'} \right| \right) dx' dy'. \tag{2.7}$$

In Eq. (2.7), $w_{x,y}^{x',y'}$ is the weight function, which indicates that the neighboring and the center pixels belong to the same surface. Usually, the weight is computed based on the spatial distance and the brightness distance between the neighboring (x', y') and the center pixel (x, y).

Zhang et al. observe that the weighted median filter relies more on the divergence of the motion flow rather than just on the brightness distance. They propose an improved method for computing the weight of pixels in a filtering window. They introduce a scale parameter λ to adjust the scale with the brightness distance in the filtering window and an occlusion factor to detect the occlusion regions of the image sequence. Despite improving the robustness of optical flow estimation, the existing weighted median filtering optimization may cause ill-defined edges or boundaries in outdoor complex conditions [24, 25]. The existing optical flow estimation methods are different in objective functions and the way of optimizing them. Some implementation details have been reported, such as the median filtering methods in the optical flow estimation [26–28].

The guided filter (GF) [29–34] is another solution to handle the problems of over-smoothing and edge-preserving in varying illumination and complex scenes. At the same time, it has also been applied in many computer vision applications, including optical flow estimation [35-37]. The GF uses a guidance image and generates the less smoothed and more structured output image. The robustness and the simplicity of GF make it beneficial for the optical flow estimation. However, it is observed that GF may produce halo artifacts near the edges because of the unsuitable smoothness of an edge.

Recently, the joint filtering-based optical flow estimation method [38–41] has been used to preserve motion boundaries. Zhang et al. [17] introduced a mutual-structure GF in a NLTV optical flow estimation framework, which is optimized using a joint filtering approach. This method successfully preserves motion boundaries while maintaining robustness against noise and illumination changes. Additionally, Li et al. [44] proposed a weighted guided image filter that incorporates edge-aware weights into the GF, allowing for better handling of halo artifacts and improving the overall robustness of the optical flow estimation process. The computational complexity of these methods remains comparable to traditional GF methods, making them feasible for real-time applications in complex outdoor environments [42, 44].

2.4 Noise and Uncertainty in Outdoor Environments

Noise refers to the random fluctuations in pixel values caused by various factors, such as imperfections in image sensors or environmental disturbances. It can manifest in different ways, from small variations in pixel intensities to larger distortions caused by changes in lighting or occlusions in the scene. Uncertainty, on the other hand, refers to the ambiguity in the motion estimation process. It arises when there is insufficient information to determine the correct motion vectors, often due to texture less regions, occlusions, or motion boundaries in the image.

Noise and uncertainty significantly impact the accuracy and reliability of motion estimates. In practical applications, motion estimation algorithms are often deployed in challenging environments, such as autonomous vehicles operating in adverse weather conditions or surveillance systems monitoring crowded scenes. In these cases, the performance of motion estimation algorithms can degrade due to noise and uncertainty, leading to errors in the estimated motion vectors. Therefore, understanding the sources of noise and uncertainty, as well as developing robust methods to handle them, is crucial for improving the performance of motion estimation algorithms in real-world scenarios.

2.4.1 Sources of Noise in Motion Estimation

Noise in motion estimation refers to the random fluctuations in pixel values that distort the true motion information in the image. There are several sources of noise that can affect the accuracy of motion estimation algorithms, ranging from sensor imperfections to environmental disturbances. In this section, we will explore the different types of noise that are commonly encountered in motion estimation tasks and discuss how they impact the accuracy of the estimated motion vectors.

2.4.2 Sensor Noise

Sensor noise arises from imperfections in the imaging sensors used to capture the image sequences. Every camera sensor has a certain level of noise, which is introduced during the process of capturing and digitizing the image. This noise can be modeled as an additive random variable that alters the true pixel intensity, leading to distortions in the captured image. Sensor noise is particularly problematic in low-light conditions, where the signal-to-noise ratio is low, and the noise becomes more prominent.

Mathematically, sensor noise can be expressed as:

$$I_{noisy}(x, y, t) = I_{true}(x, y, t) + \eta(x, y, t) \tag{2.8}$$

where $I_{noisy}(x, y, t)$ is the observed pixel intensity at location (x, y) in the image at time t, $I_{true}(x, y, t)$ is the true pixel intensity, and $\eta(x, y, t)$ is the noise term, often modeled as Gaussian noise with zero mean and a certain variance.

The presence of sensor noise can degrade the accuracy of motion estimation algorithms, as the noise introduces random variations in the pixel intensities that can be mistaken for motion. For example, in optical flow algorithms, the brightness constancy assumption assumes that the intensity of a pixel remains constant as it moves across frames. However, sensor noise violates this assumption by introducing random fluctuations in the pixel intensities, leading to inaccurate motion estimates.

Various techniques have been developed to mitigate the effects of sensor noise in motion estimation. One common approach is to apply image preprocessing techniques, such as noise filtering, before performing motion estimation. For example, bilateral filtering is a popular technique that smooths out noise while preserving edges in the image. Another approach is to use TV regularization, which penalizes large gradients in the estimated motion field, thereby reducing the influence of noise on the motion estimates.

2.4.3 Environmental Noise

Environmental noise refers to disturbances in the scene that are caused by external factors, such as changes in lighting, shadows, reflections, and occlusions. These factors introduce significant variations in the appearance of the scene, which can confuse motion estimation algorithms. For example, a sudden change in lighting conditions, such as a cloud passing over the sun, can cause large changes in pixel intensities that are unrelated to the actual motion in the scene. Similarly, occlusions, where one object in the scene blocks another, can introduce discontinuities in the motion field, making it difficult to estimate the motion of the occluded object.

In optical flow estimation, environmental noise is particularly problematic because most optical flow algorithms rely on the brightness constancy assumption, which assumes that the intensity of a pixel remains constant as it moves across frames. However, in real-world scenarios, this assumption is often violated due to environmental factors. For example, when the lighting conditions change between frames, the pixel intensities may no longer be constant, leading to errors in the estimated optical flow.

To address the effects of environmental noise, various techniques have been developed. One approach is to use photometric invariants, which are features that are invariant to changes in lighting conditions. For example, gradient-based features, such as image gradients or edge maps, are less sensitive to changes in lighting and can be used to improve the robustness of motion estimation algorithms in the presence of environmental noise. Another approach is to use shadow detection and removal techniques to eliminate the influence of shadows on the motion estimates.

2.4.4 Quantization Noise

Quantization noise arises from the process of converting continuous pixel intensity values into discrete levels during image digitization. When an image is captured by a digital camera, the continuous intensity values of the scene are mapped to a finite set of discrete intensity levels, typically represented by an 8-bit or 16-bit value. This process introduces quantization errors, as the true pixel intensity is approximated by the nearest discrete level. In regions of the image with subtle intensity changes, these quantization errors can lead to a loss of detail and introduce noise into the motion estimates.

Quantization noise is particularly problematic in motion estimation because it can obscure small, subtle motions in the image. For example, in regions of the image with low contrast, where the intensity differences between pixels are small, quantization noise can make it difficult to accurately estimate the motion. In these cases, the motion estimates may be dominated by the quantization errors, leading to inaccurate results.

To mitigate the effects of quantization noise, subpixel motion estimation techniques have been developed. These techniques estimate motion at a finer scale by interpolating between the discrete pixel values, allowing for more accurate motion estimates in regions with low contrast. For example, optical flow algorithms often use subpixel interpolation methods, such as bicubic or bilinear interpolation, to estimate the motion at subpixel accuracy, thereby reducing the influence of quantization noise.

2.4.5 Types of Uncertainty in Motion Estimation

Uncertainty in motion estimation refers to the ambiguity or lack of confidence in the estimated motion vectors. Uncertainty arises due to various factors, such as the inherent limitations of the motion estimation algorithm, the presence of noise in the image, or the complexity of the scene being captured. There are two main types of uncertainty that affect motion estimation algorithms: epistemic uncertainty and aleatoric uncertainty. In this section, we will explore these two types of uncertainty and discuss how they impact the performance of motion estimation algorithms.

2.4.6 Model Uncertainty

Model uncertainty, arises from the limitations or imperfections of the motion estimation algorithm itself. It reflects the uncertainty in the model's ability to accurately capture the true motion in the scene. Epistemic uncertainty is often caused by a lack of knowledge or information about the underlying motion patterns or by the use of an incorrect or incomplete model to describe the motion.

For example, many optical flow algorithms rely on the brightness constancy assumption, which assumes that the intensity of a pixel remains constant as it moves across frames. However, this assumption is often violated in real-world scenarios due to changes in lighting, occlusions, or motion boundaries. In these cases, the motion estimation algorithm may be unable to accurately capture the true motion, leading to epistemic uncertainty in the estimated motion vectors.

Model uncertainty can be reduced by improving the motion estimation model or by incorporating additional information into the estimation process. For example, incorporating prior knowledge about the motion patterns in the scene, such as the expected motion of objects or the geometry of the scene, can help reduce model uncertainty. Additionally, using more sophisticated motion estimation models, such as deep learning-based approaches, can improve the accuracy of the motion estimates and reduce model uncertainty.

2.4.7 Data Uncertainty

Data uncertainty, also known as data uncertainty, arises from the inherent variability or noise in the input data. It reflects the uncertainty in the estimated motion vectors due to the presence of noise, occlusions, or other disturbances in the image. Data uncertainty is often caused by factors such as sensor noise, environmental noise, or quantization errors, which introduce random variations in the pixel intensities that affect the accuracy of the motion estimates.

Unlike model uncertainty, data uncertainty cannot be reduced by improving the motion estimation model, as it is inherent in the input data. However, it can be accounted for by using uncertainty-aware motion estimation techniques that explicitly model the uncertainty in the estimated motion vectors. For example, probabilistic motion estimation models, such as Bayesian optical flow algorithms, represent the estimated motion as a probability distribution, allowing for a more accurate representation of the uncertainty in the motion estimates.

By modeling data uncertainty, motion estimation algorithms can provide not only an estimate of the motion vectors but also a measure of the confidence or uncertainty in the estimates. This information can be useful in applications where the accuracy of the motion estimates is critical, such as autonomous navigation or object tracking, as it allows the system to take uncertainty into account when making decisions based on the motion estimates.

2.5 Techniques for Robust Motion Estimation Under Noise and Uncertainty

In the presence of noise and uncertainty, traditional motion estimation techniques, such as optical flow, may produce inaccurate or unstable results. To address these challenges, various robust motion estimation techniques have been developed that are specifically designed to handle noise and uncertainty in the input data. These techniques aim to improve the accuracy and reliability of motion estimates by incorporating mechanisms to deal with noise and uncertainty in the estimation process.

2.5.1 Total Variation Regularization

TV regularization is a popular technique for robust motion estimation that aims to reduce the influence of noise by enforcing smoothness in the estimated motion field. The key idea behind TV regularization is to penalize large gradients in the motion field, which are often caused by noise or abrupt changes in pixel intensities. By imposing a smoothness constraint on the motion estimates, TV regularization helps to reduce the influence of noise and produces more stable motion estimates.

Mathematically, the TV regularization term is defined as:

$$R(u, v) = \int_{\Omega} |\nabla u| + |\nabla v| dx, \tag{2.9}$$

where u and v are the horizontal and vertical components of the motion field, ∇ denotes the gradient operator, and Ω is the image domain. The TV regularization term penalizes large gradients in the motion field, encouraging smooth transitions between motion vectors and reducing the influence of noise.

TV regularization has been widely used in optical flow algorithms to improve the robustness of motion estimation under noisy conditions. For example, the TV-L^1 optical flow algorithm incorporates a TV regularization term into the optical flow estimation process to reduce the influence of noise and produce more accurate motion estimates. The use of TV regularization has been shown to significantly improve the performance of optical flow algorithms in challenging environments, such as low-light or high-noise conditions.

2.5.2 RANSAC Algorithm

The RANSAC (Random Sample Consensus) algorithm is a robust estimation technique that is commonly used to handle outliers and noise in motion estimation tasks. The key idea behind RANSAC is to iteratively sample a subset of the data points and estimate the motion model based on this subset. The algorithm then evaluates the estimated model by checking how well it fits the remaining data points. If the model fits a large number of data points, it is considered a good estimate of the true motion. Otherwise, the algorithm repeats the process with a different subset of the data points.

RANSAC is particularly effective in scenarios where the input data contains a large number of outliers or noise, as it is able to identify and discard the outliers during the estimation process. This makes RANSAC a popular choice for robust motion estimation in applications such as feature-based optical flow, where the input data may contain noisy or incorrect feature correspondences.

In addition to its robustness to noise and outliers, RANSAC is also highly flexible and can be applied to a wide range of motion estimation tasks. For example, RANSAC can be used to estimate rigid body motion, homographies, or fundamental matrices in structure-from-motion algorithms. Its versatility and robustness make RANSAC a valuable tool for motion estimation in real-world applications.

2.6 Uncertainty-Aware Motion Estimation Methods

In addition to robust motion estimation techniques, uncertainty-aware methods have been developed to explicitly model and account for the uncertainty in the motion estimates. These methods are particularly useful in scenarios where the input data is noisy or incomplete, as they provide not only an estimate of the motion vectors but also a measure of the uncertainty in the estimates. By incorporating uncertainty into the estimation process, these methods can produce more accurate and reliable motion estimates, even in the presence of noise and uncertainty.

2.6.1 Probabilistic Motion Estimation Models

Probabilistic motion estimation models represent the estimated motion as a probability distribution, rather than a single deterministic estimate. This allows the model to capture the uncertainty in the motion estimates and provide a measure of the confidence in the estimated motion vectors. Probabilistic models are particularly useful in scenarios where the input data is noisy or incomplete, as they can account for the uncertainty in the input data and produce more robust motion estimates.

One common approach to probabilistic motion estimation is to use Bayesian inference, where the motion estimates are represented as a posterior probability distribution given the input data. For example, in Bayesian optical flow, the motion field is estimated by maximizing the posterior probability of the motion vectors given the observed image sequence. This approach allows the model to account for both the noise in the input data and the uncertainty in the motion estimation process, leading to more accurate and reliable motion estimates.

Probabilistic motion estimation models have been successfully applied to a wide range of tasks, including optical flow, 3D motion estimation, and structure-from-motion. By incorporating uncertainty into the estimation process, these models are able to handle noisy or incomplete data more effectively than traditional deterministic methods.

2.6.2 Deep Learning Approaches to Uncertainty-Aware Motion Estimation

In recent years, deep learning approaches have been increasingly used for motion estimation tasks, due to their ability to learn complex motion patterns from large datasets. Deep learning models, such as convolutional neural networks (CNNs) and recurrent neural networks (RNNs), have been shown to achieve state-of-the-art performance in motion estimation tasks, such as optical flow and video prediction.

One of the key advantages of deep learning approaches is their ability to model uncertainty in the motion estimates. By training deep learning models to predict both the motion vectors and the uncertainty in the estimates, these methods can produce more accurate and reliable motion estimates, even in challenging environments. For example, in uncertainty-aware optical flow, deep learning models are trained to predict a probability distribution over the motion vectors, rather than a single deterministic estimate. This allows the model to capture the uncertainty in the motion estimates and provide a measure of the confidence in the predictions.

Deep learning approaches to uncertainty-aware motion estimation have been applied to a wide range of tasks, including optical flow, video prediction, and 3D motion estimation. By leveraging the power of deep learning and uncertainty modeling, these methods are able to handle noisy or incomplete data more effectively than traditional motion estimation algorithms.

2.7 Real-World Applications of Noise and Uncertainty in Motion Estimation

The challenges posed by noise and uncertainty in motion estimation are not merely theoretical but have significant practical implications in a wide range of real-world applications. In this section, we will explore several applications where noise and uncertainty play a critical role in motion estimation and discuss how the techniques and methods discussed in this chapter can be applied to address these challenges.

2.7.1 Autonomous Vehicles

Autonomous vehicles rely on motion estimation algorithms to perceive and navigate their environment. These algorithms are used to estimate the motion of other vehicles, pedestrians, and objects in the scene, allowing the autonomous vehicle to plan its path and avoid collisions. However, the performance of motion estimation algorithms in autonomous vehicles is often degraded by noise and uncertainty in the input data. For example, sensor noise from cameras, lidar, or radar can introduce errors in the estimated motion vectors, while environmental factors, such as rain, fog, or low-light conditions, can further complicate the estimation process.

To address these challenges, robust and uncertainty-aware motion estimation techniques have been developed for autonomous vehicles. For example, probabilistic motion estimation models can be used to account for the uncertainty in the sensor data, while TV regularization can be applied to reduce the influence of noise on the motion estimates. These techniques enable autonomous vehicles to operate more reliably in challenging environments, where noise and uncertainty are prevalent.

2.7.2 Medical Imaging

Motion estimation plays a crucial role in many medical imaging applications, such as image registration, motion compensation, and tracking of anatomical structures. However, the accuracy of motion estimation in medical imaging is often affected by noise and uncertainty in the input data. For example, patient motion during imaging can introduce noise into the images, while variations in tissue properties can lead to uncertainty in the estimated motion of anatomical structures.

To improve the accuracy and reliability of motion estimates in medical imaging, robust and uncertainty-aware motion estimation techniques have been applied. For example, TV regularization can be used to reduce the influence of noise in medical images, while probabilistic motion estimation models can account for uncertainty in the motion estimates due to variations in tissue properties. These techniques have been shown to improve the accuracy of motion estimation in a wide range of medical imaging applications, leading to better diagnosis and treatment outcomes.

2.7.3 Augmented Reality

In AR applications, motion estimation is used to track the motion of the user's head, hands, or other objects in the scene, allowing virtual objects to be overlaid on the real-world environment. However, the accuracy of motion estimation in AR is often affected by noise and uncertainty in the input data. For example, sensor noise from cameras or inertial measurement units (IMUs) can introduce errors in the estimated motion, while variations in lighting conditions or occlusions can further complicate the estimation process.

To address these challenges, robust and uncertainty-aware motion estimation techniques have been applied to AR applications. For example, deep learning-based motion estimation models can be trained to predict both the motion vectors and the uncertainty in the estimates, allowing the AR system to account for noise and uncertainty in the input data. Additionally, TV regularization can be used to reduce the influence of noise on the motion estimates, leading to more accurate and stable AR experiences.

2.8 Handling Motion in Low-Light and Foggy Conditions

Optical flow estimation in low-light and foggy conditions presents significant challenges due to poor visibility and fluctuating lighting environments. These factors distort image features, causing a loss of vital information necessary for accurate motion detection. Traditional optical flow methods, such as Horn–Schunck and Lucas–Kanade, struggle to maintain precision in these conditions, leading to unreliable estimates of motion.

Advanced Techniques for Low-Light Scenarios: Recent advancements have introduced adaptive brightness models that dynamically adjust to varying light conditions. By employing these models, the optical flow estimation algorithms can differentiate between actual motion and noise caused by lighting inconsistencies. One example is the introduction of dynamic contrast adjustment in real-time driving simulations, which has shown to improve the detection of objects like pedestrians and vehicles in low-light environments. Moreover, attention mechanisms have become integral in motion estimation under poor lighting. These models enable the system to focus on regions of interest within the image, filtering out irrelevant background noise. For instance, in autonomous driving systems, attention-based models prioritize road signs and moving vehicles, while disregarding static objects that do not contribute to the flow. By isolating the regions of interest, the system becomes more adept at estimating motion despite environmental challenges.

Real-World Example: In night-time driving scenarios, an adaptive optical flow model utilizing attention mechanisms can significantly reduce the error rate in detecting oncoming traffic. In one case study involving traffic monitoring systems, the error rate was reduced by 25% when using adaptive brightness models compared to traditional methods.

Handling Foggy Conditions: Fog adds another layer of complexity as it obscures objects and causes scattering of light, leading to inconsistent flow vectors. In response to these challenges, researchers have developed fog-specific motion models that include haze removal techniques. These methods involve preprocessing the images with dehazing algorithms before performing optical flow calculations. For example, a foggy weather monitoring system in urban areas successfully utilized this approach, enhancing its ability to detect vehicle movement and pedestrian activity. By removing the fog, the system was able to capture clearer images, leading to more accurate motion estimates. These advances make optical flow estimation more robust in adverse conditions like fog, ensuring better performance for real-time applications such as surveillance, autonomous navigation, and environmental monitoring.

2.9 Optical Flow for Rain, Snow, and Other Weather Conditions

Rain and snow introduce unique challenges for optical flow estimation, primarily due to their dynamic nature and the noise they introduce. Rain streaks, for example, can occlude objects in the image, making it difficult to detect boundaries and edges, both of which are crucial for accurate motion estimation. Similarly, snowflakes can create a textured noise pattern that confounds traditional optical flow models, leading to significant errors in motion detection.

Decomposition Techniques: A widely adopted approach for handling rain and snow involves decomposing the scene into its constituent components: the background, motion, and weather streaks. By separating rain and snow from the main scene, optical flow algorithms can focus on the actual motion, ignoring the interference caused by weather elements. An example of this is a scene decomposition technique applied to surveillance footage during a snowstorm. By isolating the snowflakes from the scene, the algorithm could track vehicles more effectively, reducing motion detection errors by up to 30%. This technique is particularly valuable in outdoor surveillance and autonomous driving systems where weather interference can lead to false detections or missed objects.

Recent Advances: Incorporating deep learning models has significantly improved the performance of optical flow under rainy and snowy conditions. One notable approach uses CNNs to learn weather-specific noise patterns, allowing for automatic adjustment of the motion estimation process. For instance, in outdoor surveillance systems, CNN-based models that learned from rain and snow conditions were able to adapt their optical flow estimation in real time, improving the detection of moving objects even in heavy precipitation.

2.10 Deep Learning-based Optical Flow Estimation

Deep learning techniques, particularly CNNs, have gained prominence in optical flow estimation due to their ability to learn complex motion patterns from large datasets. These methods outperform classical techniques like Horn–Schunck by automatically identifying relevant features, such as edges and motion boundaries, without the need for hand-crafted regularization terms.

2.10.1 FlowNet Architecture

FlowNet is a pioneering neural network model for optical flow estimation, developed to directly learn flow patterns from large datasets. Unlike traditional methods that rely on handcrafted regularization, FlowNet's architecture is based on an encoder–decoder structure with skip connections, enabling it to capture both local and global motion.

The original FlowNet model was improved with the development of FlowNet2, which incorporates multiple stages of flow refinement. Each stage computes a residual flow, refining the flow estimation from the previous stage. FlowNet2 has shown substantial improvements in handling complex motion patterns and outperforms classical methods in real-world scenarios like video game animation and visual effects.

2.10.2 PWC-Net: A Pyramid and Warping Approach

PWC-Net introduced several innovations to optical flow estimation, including pyramidal processing and cost volume warping. The pyramidal structure allows for multi-scale flow estimation, improving both accuracy and computational efficiency. Cost volume warping, another key feature, allows the network to align features between successive frames, thereby improving motion estimation.

In real-time applications such as autonomous driving, PWC-Net has proven to be more efficient than traditional methods, offering both high accuracy and reduced computational complexity. For example, a study in urban traffic management demonstrated that PWC-Net could estimate vehicle motion in real time, outperforming classical techniques by 15% in terms of accuracy while reducing computational overhead by 30%.

2.11 Optical Flow with Recurrent Neural Networks

RNNs have brought a new dimension to optical flow estimation, particularly in their ability to capture temporal dependencies in video sequences. Unlike traditional feedforward networks, RNNs are designed to process sequential data, making them highly effective for tasks where the temporal relationship between frames is critical. One of the most common types of RNNs applied in optical flow is the long short-term memory (LSTM) network, which is designed to remember long-term dependencies in time-series data. This makes RNNs particularly useful for estimating motion over extended video frames, handling occlusions, and predicting future motion in dynamic environments.

2.11.1 LSTM-based Optical Flow

LSTM-based models excel in situations where optical flow estimation requires a memory of previous frames to handle occlusions, ambiguous motion, or long-range temporal dependencies. By learning from sequences of frames, these networks can estimate motion even in regions that are occluded or where motion is complex, such as rapidly changing scenes.

Sports Analytics: In sports analytics, capturing motion accurately across time is crucial for tracking player movement, strategy analysis, and injury prevention. Consider a scenario in basketball, where an LSTM-based optical flow model is applied to a sequence of game footage to track the movement of players and the ball. The LSTM model, trained on temporal sequences, can accurately estimate the flow even when players momentarily occlude each other or the ball, which would be difficult for traditional optical flow methods to handle. The model "remembers" the motion patterns from earlier frames, allowing it to predict motion more accurately in later frames, ensuring that even fast, occluded motions like a ball passing through a crowd are captured.

Autonomous Driving in Traffic: In autonomous driving, RNNs and LSTM-based optical flow systems are utilized to track vehicles, pedestrians, and other objects over time, enabling the vehicle to predict future motion and react accordingly. In a scenario where a car is driving through busy traffic, the LSTM-based optical flow model can maintain a continuous estimate of the motion of surrounding vehicles. Even if a vehicle temporarily moves out of the field of view or is occluded by another car, the model retains knowledge of its previous motion, improving its ability to predict the vehicle's reappearance. This temporal coherence allows for smoother and more reliable decision-making in dynamic, real-world environments.

2.11.2 Capturing Temporal Consistency in Optical Flow

A significant advantage of RNNs, especially LSTMs, is their ability to maintain temporal consistency across frames, which is critical for many applications where motion evolves gradually or involves repeated patterns. Traditional optical flow models often struggle with temporal noise or inconsistencies, especially in video sequences where lighting or motion may change abruptly. LSTM-based models, however, utilize their memory cells to learn long-range dependencies, making them better suited for these situations.

Surveillance in Crowded Environments: In crowded environments like airports or shopping malls, surveillance systems need to track multiple people over long periods, often with overlapping motion. An LSTM-based optical flow system applied to such surveillance footage can accurately capture the motion of each individual even when they temporarily move out of frame or are occluded by others. The recurrent nature of the LSTM ensures that motion patterns are retained, allowing for a more consistent tracking of individuals over time. This is particularly useful in security applications where reliable identification and motion tracking are paramount.

Action Recognition in Videos: In the context of action recognition, RNNs combined with optical flow can be used to classify complex movements in video sequences. For example, in a video of a person performing martial arts, an LSTM-based optical flow model can track the flow of body movements across multiple frames, capturing the subtle temporal shifts in motion that define each action. By leveraging both spatial and temporal data, the system can distinguish between different actions, such as kicking, punching, or jumping, even when the motion is fast or partially occluded.

2.11.3 Case Study: LSTM Optical Flow for Pedestrian Tracking

In a study involving pedestrian tracking for autonomous systems, an LSTM-based optical flow model was deployed to track pedestrians over multiple frames of video captured in an urban environment. The model learned the movement patterns of pedestrians, including sudden stops, turns, and occlusions behind obstacles such as vehicles or trees. Compared to traditional optical flow techniques, the LSTM-based model achieved higher accuracy in predicting future pedestrian positions, reducing errors in both crowded and sparsely populated scenes.

Traffic Monitoring Systems: In traffic monitoring systems, accurately estimating the movement of vehicles over time is crucial for congestion analysis and accident prevention. LSTM-based optical flow models are applied to continuous video feeds from traffic cameras, learning the motion patterns of vehicles over time. The recurrent nature of the LSTM allows the model to maintain a consistent estimate of traffic flow even when visibility is poor due to weather conditions, or vehicles are briefly occluded by others. This improves the reliability of traffic predictions and assists in dynamic

2.11.4 Handling Occlusions and Temporal Ambiguities

Occlusions and temporal ambiguities present significant challenges for motion estimation in dynamic environments. LSTM-based optical flow models are particularly effective at handling such scenarios, as they can "remember" the flow from previous frames, allowing them to estimate motion even in regions that become temporarily invisible or ambiguous.

Video Editing for Special Effects: In video editing, where special effects often rely on tracking motion over time, LSTM-based optical flow models can ensure that motion remains consistent even when objects or actors are temporarily occluded by other elements in the scene. For instance, in a scene where a character moves behind an object and reappears, the LSTM model can predict the character's movement trajectory, allowing editors to seamlessly apply effects like motion blur or animation without losing track of the motion.

2.11.5 Future Directions: Combining LSTMs with 3D Optical Flow

As optical flow extends into 3D vision, integrating LSTM-based models with multi-view and depth-aware systems represents a promising future direction. By combining the temporal consistency of LSTMs with the spatial accuracy of 3D optical flow, it will be possible to achieve more reliable and precise motion estimation for applications such as virtual reality (VR), AR, and autonomous systems. This integration will allow systems to track motion across both time and space, enabling richer and more immersive experiences in a wide range of applications.

2.12 Self-Supervised and Unsupervised Learning Approaches

One of the limitations of supervised learning approaches in optical flow estimation is the need for large amounts of labeled ground-truth data. Self-supervised and unsupervised techniques have emerged as effective alternatives, allowing networks to learn motion patterns without explicit labels. This section explores these approaches, including methods that rely on image reconstruction and photometric consistency losses. A common approach in unsupervised optical flow estimation is the use of photometric loss, which ensures that the predicted flow aligns with the intensity changes between consecutive frames. This allows for the learning of motion directly from the data, bypassing the need for labeled ground truth. In real-world applications, unsupervised methods have shown remarkable success in tasks such as autonomous navigation, where manually labeling large datasets is impractical. For instance, in self-driving car systems, unsupervised learning techniques have enabled real-time motion estimation without requiring extensive labeled datasets.

2.13 Optical Flow in 3D Vision

Optical flow has traditionally been applied in 2D, but recent advancements have extended its use to 3D vision. Multi-view systems can estimate optical flow across multiple cameras, providing more accurate motion estimation for applications such as VR and AR.

2.13.1 Multi-view Optical Flow Estimation

In multi-camera setups, optical flow can be computed across different camera angles to achieve a more accurate representation of motion in 3D space. This approach has been used extensively in applications like VR and AR, where accurate motion estimation is crucial for creating immersive experiences. For example, in 3D animation, multi-view optical flow estimation has been used to capture complex motions from different angles, allowing animators to create more realistic characters and environments. By enhancing optical flow estimation through these advanced techniques and leveraging deep learning methods, the field is rapidly evolving to meet the challenges of real-world applications, from low-light conditions to multi-view 3D motion analysis

2.14 Optical Flow Estimation-based Motion Segmentation

Motion segmentation plays a critical role in the field of computer vision by grouping pixels according to different motion patterns exhibited by independently moving objects. It involves the task of separating different moving objects within a video or image sequence by analyzing the motion vectors at each pixel. This process becomes significantly easier when the motion flow of each pixel is known. However, motion segmentation and optical flow estimation, despite being closely related, are often treated as two separate tasks in traditional computer vision pipelines. In recent years, there have been growing efforts to combine both tasks into a single cohesive framework, which not only enhances segmentation accuracy but also provides better estimates of object boundaries. In this section, we provide an in-depth review of traditional and recent motion segmentation methods that leverage optical flow for more accurate and robust segmentation.

2.14.1 Traditional Motion Segmentation Approaches

Motion segmentation groups the pixels according to different motion patterns of independently moving objects [45–47]. If we know the motion flow of each pixel, the task of segmenting different moving objects becomes easier. Optical flow estimation and motion segmentation are two important tasks in computer vision, but they are often handled separately. Optical flow methods estimate the motion field in an image sequence by assuming that the intensity of moving objects remains constant over time. On the other hand, motion segmentation aims to partition the image into regions corresponding to different objects based on their motion. This separation of tasks has historically limited the performance of both tasks, especially in scenarios with complex motion patterns or adverse conditions. Next, we will review some of the relevant works that aim to combine these two tasks more effectively.

2.14.2 Limitations of Traditional Optical Flow Estimation-based Segmentation Methods

Most optical flow estimation-based segmentation methods are sensitive to illumination changes due to the brightness constancy assumption. This assumption implies that pixel intensities do not change as objects move, but in real-world conditions, changes in lighting, shadows, and reflections can violate this assumption, leading to errors in motion estimation. Moreover, they may suffer from boundary blurring problems. For example, Chen et al. [50] propose a dense optical flow estimation method for dynamic natural phenomena (i.e., water, fire, etc.) and use the estimated motion field to segment the phenomena. Nevertheless, the method can only estimate and segment the natural phenomena instead of a general object motion.

2.14.3 Layer-based Motion Segmentation Methods

To overcome some of the challenges associated with traditional optical flow-based methods, several layer-based approaches have been proposed in recent years. These methods represent the scene as a collection of motion layers, where each layer corresponds to a different moving object. Motion layers provide a more flexible representation of the scene by allowing objects to be segmented based on their unique motion characteristics. These approaches have shown robustness for motion segmentation, particularly in controlled environments where the motion field is assumed to have already been estimated and can be directly used for segmentation.

In layer-based motion segmentation, the combination of optical flow estimation and alpha matting within a graph cut framework is a popular technique for achieving robust motion layer segmentation [49, 50]. These methods attempt to segment the scene into layers by minimizing an energy function that takes into account both the optical flow and the alpha matte, which describes the transparency of each pixel. The graph cut optimization algorithm is then used to find the optimal segmentation of the scene. One key limitation of these approaches is that they tend to perform well only in simple scenarios with a small number of moving objects. As the number of objects increases or the scene becomes more complex, the performance of these methods degrades due to the increased difficulty of estimating multiple motion layers simultaneously [51].

The low-contrast problem is another significant challenge for traditional layer-based motion segmentation methods. This problem occurs when the moving objects in the scene have similar intensities or colors to the background, making it difficult to distinguish them based on optical flow alone. To address this issue, some methods rely on background modeling techniques, where a dynamic matrix is used to represent the appearance and motion of the background. Moving objects are then segmented by detecting deviations from the background model [52]. However, these methods are highly sensitive to noisy environments and adverse outdoor conditions, such as rain, fog, or snow, where the background model can become unreliable. Noisy environments often introduce spurious motion artifacts into the optical flow field, which can lead to inaccurate segmentation of moving objects. Therefore, designing robust optical flow estimation and motion segmentation methods that can perform well under such conditions has been a less-studied area [53, 54].

2.14.4 Deep Learning for Joint Optical Flow and Motion Segmentation

In recent years, deep learning techniques have emerged as powerful tools for addressing the limitations of traditional motion segmentation approaches. One of the key advantages of deep learning-based methods is their ability to learn complex representations of motion and appearance directly from the data, without relying on hand-crafted features or prior assumptions about the scene. This has led to the development of models that can jointly estimate optical flow and segment moving objects in a single pipeline.

Recent deep learning techniques focus on combining optical flow estimation and motion segmentation into a single pipeline, allowing for more accurate and simultaneous prediction of motion and object boundaries. This section reviews notable works such as MaskFlowNet, which uses flow information to guide object segmentation. MaskFlowNet is designed to predict both the optical flow field and the motion segmentation mask in a unified architecture, enabling it to handle complex scenes with multiple moving objects and varying lighting conditions. The use of deep learning also allows for the incorporation of large-scale datasets, enabling the model to generalize better to real-world scenarios where traditional methods struggle.

By leveraging the power of CNNs and RNNs, these models can capture both the spatial and temporal aspects of motion, leading to more robust segmentation results in dynamic scenes. For example, in scenarios where objects occlude each other or move in complex patterns, deep learning-based approaches can maintain accurate motion segmentation by learning from large amounts of training data. Additionally, these methods are less sensitive to illumination changes, occlusions, and noise, which are common challenges in traditional optical flow estimation methods.

2.15 Evaluation Measures

In this section, we analyze two types of evaluation measures to evaluate the performance of optical flow estimation methods.

2.15.1 Quantitative Evaluation

Two commonly used mathematical measures for the optical flow estimation methods are the Average Angular Error (AAE) and the End-Point Error (EPE).

Average Angular Error (AAE): The AAE is the angle between the estimated flow vectors $(\hat{u}_{est}, \hat{v}_{est})$ and the ground-truth flow vectors (u_{grt}, v_{grt}) as follows:

$$AAE = \cos^{-1}\left(-\frac{\hat{u}_{est}, u_{grt} + \hat{v}_{est}, v_{grt} + 1}{\sqrt{1 + \hat{u}_{est}^2 + \hat{v}_{est}^2}\sqrt{1 + u_{grt}^2 + v_{grt}^2}} \right). \tag{2.10}$$

End-Point Error (EPE): The EPE is the Euclidean distance between the estimated flow vectors and the ground-truth flow vectors, and it is defined as follows:

$$EPE = \sqrt{(\hat{u}_{est} - u_{grt})^2 + (\hat{v}_{est} - v_{grt})^2}. \tag{2.11}$$

2.15.2 Qualitative Evaluation

In addition to quantitative measures, qualitative evaluation plays an essential role in assessing the performance of optical flow estimation methods under real-world conditions. This is particularly important in applications where the visual quality of the motion segmentation is more critical than numerical accuracy, such as in video editing or animation. Several challenging datasets have been established to evaluate the optical flow estimation methods qualitatively. However, it is more challenging to evaluate the robustness of the optical flow estimation methods under real-world outdoor complex conditions. In this section, we review both the most commonly used datasets for optical flow estimation methods and the datasets of complex outdoor scenes.

2.16 Datasets

The availability of large-scale datasets has been instrumental in advancing the development of optical flow estimation and motion segmentation methods. These datasets provide ground-truth annotations that allow researchers to train and evaluate their models on a wide range of real-world scenarios. Some of the most widely used datasets for optical flow estimation and motion segmentation include the Middlebury dataset, MPI-Sintel dataset, Foggy Zurich dataset, and CDnet2014 dataset.

The Middlebury dataset [55] is the most widely used dataset for evaluating the optical flow estimation methods, and it supplies the ground truths for both real and synthetic image sequences. It contains complex conditions such as nonrigid motion, sensor noise, and motion discontinuities. The image sequences are captured in a controlled lab environment. The synthetic image sequences are generated using computer graphic techniques. However, the Middlebury is a small-scale dataset, and it is not designed to provide training data for machine learning-based optical flow estimation methods. The resolution of the Middlebury dataset images ranges from 548×388 to 640×480.

The MPI-Sintel dataset [56] is based on an open-source animated short film (i.e., "Sintel"). It contains complex conditions such as large motions, blur effects, and shading. The image sequences are rendered with complicated and challenging scenarios for current optical flow estimation methods. However, these image sequences are less realistic for outdoor surveillance applications. Most image sequences are 50 frames long, and they have the ground-truth flow field. The MPI-Sintel dataset provides 1628 image frames of the ground-truth flow in a test set (which includes 564 image frames) and a training set (which includes 1064 image frames).

The abovementioned datasets are challenging. However, it is more challenging to evaluate the robustness of the optical flow estimation methods in real-world outdoor adverse scenarios. Therefore, we also evaluate our proposed methods on more real-world datasets (i.e., the Foggy Zurich [57], the Traffic [58], and the CDnet2014 [59] datasets), which are introduced as follows.

The Foggy Zurich dataset consists of a foggy cityscape sequence and a foggy driving sequence. Each sequence consists of more than 100 color images depicting real-world misty scenes. The maximum image resolution is 768×576.

The Traffic dataset contains six sequences, and most of them have more than 1000 image frames. This is a road traffic dataset obtained by a surveillance camera and has different complex conditions (e.g., fog, snow, normal conditions, etc.). The size of the image frames is 768×576 pixels.

The CDnet2014 dataset contains various video categories (e.g., challenging weather, night, shadow, etc.). Each category has four to six image sequences. There are 53 different sequences in total. The resolution of image frames varies from 320×240 to 720×576 pixels. Moreover, each video sequence contains $900 - 7000$ image frames. All video sequences have different challenging scenarios, which make this dataset appropriate for evaluating the robustness of the optical flow estimation methods.

2.17 Conclusion

In this chapter, we have provided a comprehensive review of optical flow estimation-based motion segmentation methods. We have discussed the limitations of traditional approaches, the advantages of layer-based methods, and the recent advancements in deep learning techniques that combine optical flow estimation and motion segmentation into a single pipeline. We have also reviewed the most widely used datasets for evaluating optical flow estimation and motion segmentation methods, highlighting the importance of both quantitative and qualitative evaluation measures. In the next chapter, we will delve deeper into robust optical flow estimation methods that are specifically designed to handle adverse outdoor conditions, such as rain, fog, and snow.

Examples with Solutions

Aperture Problem

Problems:
Consider a moving object with a horizontal stripe pattern (parallel lines). The object moves horizontally to the right by 5 pixels between two frames. Explain why an optical flow algorithm using only local gradients fails to detect this motion.

Step-by-Step Solution:

1) **Setup:**
   - Within a small aperture (e.g., a 5×5 window), the gradient is constant only in the vertical direction (perpendicular to the stripes).
   - The gradient in the horizontal direction = 0.
2) **Why problem occurs:**
   - Optical flow uses:

     $$I_x u + I_y v + I_t = 0.$$

     Here, $I_x = 0$ (no horizontal gradient), so the equation reduces to $I_y v + I_t = 0$.
   - Only the vertical motion (v) is constrained. Horizontal motion (u) remains ambiguous.
3) **Implication:**
   - The algorithm cannot "see" horizontal displacement inside the aperture.
   - It produces wrong or incomplete flow vectors.
4) **Solution:**
   - Add smoothness constraints across neighboring pixels → neighbors with texture provide horizontal information.
   - This resolves ambiguity by propagating reliable information into ambiguous regions.

Learning Point:
The aperture problem shows why local information is insufficient and why smoothness constraints are critical.

Brightness Constancy Assumption

Problem:
A pixel at (100, 150) has intensity 120 in frame 1. In frame 2, the same object moves right by 2 pixels, but due to illumination change, intensity increases to 130. Can the brightness constancy assumption hold here?

Solution:

1) Brightness constancy states:

$$\mathrm{I}(x, y, t) = I(x + u, y + v, t + 1).$$

   Substituting:
   - Left side = 120
   - Right side = 130
2) **Clearly:** $120 \neq 130 \rightarrow$ assumption violated.
3) **Impact on algorithms:**
   - Optical flow will interpret brightness increase as motion error.
   - Wrong motion vector may be produced.
4) **Workaround:**
   - Use gradient-based or illumination-invariant features.
   - Or replace L^2 norm with L^1 norm (TV-L^1 method) to reduce sensitivity.

Learning Point:
Brightness constancy is fragile under illumination change → robust extensions are needed in real world.

Horn–Schunck Variational Method

Problem:
Estimate the optical flow between two frames where a block moves right by 1 pixel. Assume brightness constancy holds and gradients are:

$$I_x = 2, I_y = 1, I_t = -2.$$

Use Horn–Schunck energy minimization.

Solution:

1) Energy function:

$$E(u, v) = (I_x u + I_y v + I_t)^2 + \lambda(||\nabla u||^2 + ||\nabla v||^2)$$

2) Plug in values (ignoring smoothness for his local example):

$$(2_u + 1_v - 2)^2$$

3) Solve constraint equation:

$$2_u + 1_v - 2$$

4) Infinite solutions exist (aperture problem!). Example:
   - If $u = 1, v = 0 \rightarrow$ valid solution.
   - If $u = 0.5, v = 1 \rightarrow$ also valid.

5) Smoothness assumption resolves by enforcing neighboring pixels' motion similarity.

Learning Point:
Without smoothness, flow is underdetermined. Horn–Schunck solves this by combining brightness constancy with global smoothness.

TV-L^1 versus L^2

Problem:
Suppose brightness change introduces an error of 20 intensity units. Compare penalty in L^2 versus L^1 norm.

Solution:

1) **L^2 penalty:**
   $20^2 = 400$ (error magnified heavily).
2) **L^1 penalty:**
   $|20| = 20$ (linear error, less sensitive).
3) **Implication:**
   - L^2 norm exaggerates big errors → bad for illumination changes.
   - L^1 norm is more robust → preferred in TV-L^1 method.

Learning Point:
TV-L^1 is better for real-world noisy data because it reduces outlier influence.

Noise and Quantization

Problem:
A pixel's true intensity = 125. Due to sensor noise, observed values = 123, 126, 127 in different frames. How does this affect optical flow?

Solution:

1) Brightness constancy assumption broken → each noisy intensity suggests different motion.
2) Flow algorithm may estimate false "micro-motion".
3) Use TV regularization or bilateral filtering to smooth noise while preserving edges.

Learning Point:
Preprocessing and robust regularization are key to handling noise.

Lucas–Kanade Local Method

Problem:
Given a 3×3 window with gradients:

- Average $I_x = 1.5$,
- Average $I_y = 0.5$,
- Average $I_t = -2$.

Use the Lucas–Kanade method to estimate optical flow (u, v).

Solution:

1) Lucas–Kanade solves:

$$A \begin{bmatrix} u \\ v \end{bmatrix} = -b$$

where

$$A = \begin{bmatrix} I_x^2 & I_x I_y \\ I_x I_y & I_y^2 \end{bmatrix}, b = \begin{bmatrix} I_x I_t \\ I_y I_t \end{bmatrix}.$$

2) Compute A:

$$A = \begin{bmatrix} 1.5^2 & (1.5)(0.5) \\ (1.5)(0.5) & 0.5^2 \end{bmatrix} = \begin{bmatrix} 2.25 & 0.75 \\ 0.75 & 0.25 \end{bmatrix}.$$

3) Compute b:

$$b = \begin{bmatrix} (1.5)(-2) \\ (0.5)(-2) \end{bmatrix} = \begin{bmatrix} -3 \\ -1 \end{bmatrix}.$$

4) Solve:

$$\begin{bmatrix} u \\ v \end{bmatrix} = -A^{-1} b$$

Determinant = (2.25)(0.25) – (0.75)(0.75) = 0.

Learning Point:
Lucas–Kanade fails when gradients are not varied enough (low texture). Needs good features to track.

Multi-scale Optical Flow

Problem:
An object moves by 16 pixels between two frames (too large for local gradient-based methods). How can multi-scale (pyramidal) methods estimate its flow?

Solution:

1) Build image pyramid (reduce image resolution).
2) At coarse level: 16-pixel motion looks like only 2–3 pixels. Easier to track.
3) Estimate flow at coarse level, then propagate down to finer levels.
4) Refine motion at each level until reaching original resolution.

Learning Point:
Multi-scale methods handle large motions efficiently.

Occlusion Case

Problem:
In a video, a car passes behind a tree. Pixels corresponding to the car disappear suddenly (occlusion). What happens to optical flow?

Solution:

1) Brightness constancy assumption fails (object disappears).
2) Flow algorithms give wrong/unreliable vectors at occluded pixels.
3) Solutions:
   - Use occlusion detection masks.
   - Combine flow with segmentation to ignore occluded regions.

Learning Point:
Classical methods don't handle occlusions well → later models add occlusion-aware terms.

Small Motion Approximation

Problem:
Why do classical models assume small motion?

Solution:

1) They rely on Taylor series expansion:

$$I(x+u, y+v, t+1) \approx I(x, y, t) + I_x u + I_Y v + I_t.$$

2) This expansion is valid only if u, v are small.
3) If displacement is large (say 10 pixels), higher-order terms ignored → large errors.
4) Fix: Use image pyramids or iterative refinement.

Learning Point:
Small motion assumption simplifies equations but fails in real-world fast motions.

Brightness Change versus Motion

Problem:

Frame 1: Pixel intensity = 100.
Frame 2: Pixel intensity = 120, but no movement.

How will classical optical flow interpret this?

Solution:

1) Classical model assumes: change in intensity = due to motion.
2) It falsely generates motion vectors.
3) This is a **false positive** caused by illumination change.
4) Robust fixes: $TV\text{-}L^1$, normalized gradients, feature-based flow.

Learning Point:
Intensity $\neq$ motion $\rightarrow$ must account for lighting changes.

Edge Motion Ambiguity

Problem:
An edge (vertical line) moves right by 5 pixels. Why is motion along the edge detected but not perpendicular?

Solution:

1) Gradient is only in the horizontal direction (normal to edge).
2) Flow equation constrains motion only in that direction.
3) Motion along edge direction remains ambiguous.
4) Fix: Use smoothness constraints or multiple edges crossing.

Learning Point:
Edges cause directional ambiguity in optical flow.

Comparing Horn–Schunck and Lucas–Kanade

Problem:
Why does Horn–Schunck often produce smoother results than Lucas–Kanade?

Solution:

1) **Horn–Schunck:** Global method, enforces smoothness everywhere. Produces continuous flow fields.
2) **Lucas–Kanade:** Local method, computes per window. No global smoothness.

3) In noisy regions, Lucas–Kanade may break down; Horn–Schunck propagates reliable information.

Learning Point:
Horn–Schunck = global smoothness; Lucas–Kanade = local precision.

References

1 Kunz, M., Ostrowski, A., and Zipf, P. (2014). An fpga-optimized architecture of horn and schunck optical flow algorithm for real-time applications. 1–4.

2 Navarro, J. and Buades, A. (2017). Robust and dense depth estimation for light field images. *IEEE Transactions on Image Processing* 26 (4): 1873–1886.

3 Sevilla-Lara, L., Sun, D., Learned-Miller, E.G., and Black, M.J. (2014). Optical flow estimation with channel constancy. *European Conference on Computer Vision,* Switzerland, September 6–12, 2014. 423–438.

4 Dawkins, M.S., Wang, L., Ellwood, S.A. et al. (2021). Optical flow, behaviour and broiler chicken welfare in the uk and switzerland. *Applied Animal Behaviour Science* 234 (1): 105180–105190.

5 Kharbat, M., Aouf, N., Tsourdos, A., and White, B. (2008). Robust brightness description for computing optical flow. *British Machine Vision Conference,* 1st–4th September 2008, The University of Leeds. 1–10.

6 Zhou, J. and Kwan, C. (2018). Anomaly detection in low quality traffic monitoring videos using optical flow. *Pattern Recognition and Tracking XXlX,* 122–132.

7 Sun, W., Sun, M., Zhang, X. et al. (2020). Moving vehicle detection and tracking based on optical flow method and immune particle filter under complex transportation environments. *Complexity* 2020 (1): 1–16.

8 Weickert, J., Schnorr, C., and Schnorr, S. (2001). Variational optic flow computation with a spatio-temporal smoothness constraint. *Journal of Mathematical Imaging and Vision* 14 (3): 245–255.

9 Ren, Z., Gallo, O., Sun, D. et al. (2019). A fusion approach for multi-frame optical flow estimation. *IEEE Winter Conference on Applications of Computer Vision,* Jan. 7 2019 to Jan. 11 2019. Waikoloa Village, HI, USA. 2077–2086.

10 Khan, M. and Kumar, P. (2021). Discontinuity preserving optical flow based on anisotropic operator. *International Conference on Information & Communication Technology and System,* 20th to 21st October 2021 Online. 306–311.

11 Mohamed, M.A., Rashwan, H.A., Mertsching, B. et al. (2014). Illumination-robust optical flow using a local directional pattern. *IEEE Transactions on Circuits and Systems for Video Technology* 24 (9): 1499–1508.

12 Lejmi, W., Mahjoub, M.A., and Ben, A.K. (2017). Event detection in video sequences: challenges and perspectives. *International Conference on Natural Computation, Fuzzy Systems and Knowledge Discovery*, 29 July 2017 – 31 July 2017 Conference Center – Guilin University of Electronic Technology. 682–690.

13 Kashif, R. and Rao, S. (2022). Optical flow estimation technique for hazy scenes. *International Conference on Digital Futures and Transformative Technologies,* 24–26 May 2022, Rawalpindi, Pakistan. 1–6.

14 Wu, M., Cao, X., Cao, X., and Guo, S.A. (2022). Dataset of ant colonies motion trajectories in indoor and outdoor scenes for social cluster behavior study. https://cn.depositphotos.com/home.html.

15 Lee, H., Homeyer, C., Herzog, R. et al. (2022). Spatio-temporal outdoor lighting aggregation on image sequences using transformer networker, arXiv preprint: 2202.09206.

16 Li, R., Tan, R.T., and Cheong, L.F. (2018). Robust optical flow in rainy scenes. *European Conference on Computer Vision,* Munich, Germany, September 8–14. 288–304.

17 Zhang, C., Ge, L., Chen, Z. et al. (2019). Refined tv-l1 optical flow estimation using joint filtering. *IEEE Transactions on Multimedia* 22 (2): 349–364.

18 Inkawhich, N., Inkawhich, M., Chen, Y., and Li, H. (2018). Adversarial attacks for optical flow-based action recognition classifiers, arXiv preprint: 1811.11875.

19 Kumar, P. (2020). A duality based approach for fractional order tv-model in optical flow estimation. *IEEE International Conference on Computing, Communication and Security,* Dec. 29 2020 to Jan. 1 2021 Guangzhou, China. 1–5.

20 Thanh, D.N.H., Thanh, L.T., Hien, N.N., and Prasath, S. (2020). Adaptive total variation l1 regularization for salt and pepper image denoising. *Optik* 208 (1): 163677–163687.

21 Wedel, A., Pock, T., Zach, C. et al. (2009). An improved algorithm for tv-l1 optical flow. Statistical and geometrical approaches to visual motion. *Analysis,* 23–45.

22 Guo, X., Li, S., Li, L., and Zhang, J. (2018). Structure-texture decomposition via joint structure discovery and texture smoothing. *IEEE International Conference on Multimedia and Expo,* July 23 2018 to July 27 2018. San Diego, CA, USA. 1–6.

23 Hao, S., Han, X., Guo, Y. et al. (2020). Low-light image enhancement with semi-decoupled decomposition. *IEEE Transactions on Multimedia* 22 (12): 3025–3038.

24 Hur, J. and Roth, S. (2019). Iterative residual refinement for joint optical flow and occlusion Estimatio. *IEEE Computer Society Conference on Computer Vision and Pattern Recognition,* June 15 2019 to June 20 2019 Long Beach, CA, USA. 5754–5763.

25 Yalavarthy, P.K., Kalva, S.K., Pramanik, M., and Prakash, J. (2021). Non-local means improves total-variation constrained photoacoustic image reconstruction. *Journal of Biophotonics* 14 (1): 1–8.

26 Smitha, A., Febin, I.P., and Jidesh, P. (2022). A retinex based non-local total generalized variation framework for oct image restoration. *Biomedical Signal Processing and Control* 71 (1): 103234–103245.

27 Sun, D., Roth, S., and Black, M.J. (2014). A quantitative analysis of current practices in optical flow estimation and the principles behind them. *International Journal of Computer Vision* 106 (1): 115–137.

28 Alqadi, Z.A. and Barakat, M.T. (2021). A case study to improve the quality of median filter. *International Computer Science Mobile Computing* 10 (11): 19–28.

29 George, G., Oommen, R.M., Shelly, S. et al. (2018). A survey on various median filtering techniques for removal of impulse noise from digital image. *Conference on Emerging Devices and Smart Systems*, 235–238. 2–3 March 2018, Tiruchengode, India.

30 Chen, J., Cai, Z., Lai, J., and Xie, X. (2018). A filtering-based framework for optical flow estimation. *IEEE Transactions on Circuits and Systems for Video Technology* 29 (5): 1350–1364.

31 Zheng, Y., Zhang, M., and Lu, F. (2020). Optical flow estimation in the dark. *IEEE Conference on Computer Vision and Pattern Recognition*, 6749–6757. June 13 2020 to June 19 2020. Seattle, WA, USA.

32 Cho, J., Jung, Y., Kim, D.S. et al. (2019). Moving object detection based on optical flow estimation and a gaussian mixture model for advanced driver assistance systems. *Sensors* 19 (14): 3217–3227.

33 Li, Y. and Osher, S. (2009). A new median formula with applications to pde based denoising. *Communications in Mathematical Sciences* 7 (3): 741–753.

34 Tomasi, C. and Manduchi, R. (1998). Bilateral filtering for gray and color images. *IEEE International Conference on Computer Vision,* 839–846. Jan. 4 1998 to Jan. 7 1998. Bombay, India.

35 Liu, W., Chen, X., Shen, C. et al. (2017). Robust guided image filtering, arXiv preprint: 1703.09379.

36 Qiu, X., Li, M., Zhang, L., and Yuan, X. (2019). Guided filter-based multi-focus image fusion through focus region detection[J]. *Signal Processing: Image Communication* 72 (1): 35–46.

37 Hosni, A., Bleyeri, M., Rhemann, C. et al. (2011). Real-time local stereo matching using guided image filtering. *IEEE International Conference on Multimedia and Expo,* 1–6. July 11 2011 to July 15 2011. Barcelona.

38 Wu, J., Wang, X., Chen, Z., and Zhang, C. (2018). Robust variational optical flow algorithm based on rolling guided filtering. *International Congress on Image and Signal Processing, BioMedical Engineering and Informatics*, 1–6. Oct 13–15, 2018, Beijing, China.

39 Tu, Z., Poppe, R., and Veltkamp, R.C. (2016). Adaptive guided image filter for warping in variational optical flow computation. *Signal Processing* 127 (1): 253–265.

40 Liu, N. and Zhao, D. (2014). Detail enhancement for high-dynamic-range infrared images based on guided image filter. *Infrared Physics & Technology* 67 (1): 138–147.

41 Shen, X., Zhou, C., Xu, L., Jia, J. (2015). Mutual-structure for joint filtering. *IEEE Inernational Conference on Computer Vision,* 3406–3414. Dec. 7 2015 to Dec. 13 2015. Santiago, Chile.

42 Stoll, M., Volz, S., and Bruhn, A. (2013). Joint trilateral filtering for multiframe optical flow. *IEEE International Conference on Image Processing,* 3845–3849. Melbourne, Australia, September 15–18, 2013.

43 Kim, B., Ponce, J., and Bumsub, H. (2021). Deformable kernel networks for joint image filtering. *International Journal of Computer Vision* 129 (2): 579–600.

44 Li, Z., Zheng, J., Zhu, Z. et al. (2014). Weighted guided image filtering. *IEEE Transactions on Image Processing* 24 (1): 120–129.

45 Rao, S. and Wang, H. (2021). Robust optical flow estimation via edge preserving filtering. *Signal Processing: Image Communication* 96 (1): 116295–116309.

46 Rao, S. and Wang, H. (2022). Optical flow estimation via weighted guided filtering with non-local steering kernel. *The Visual Computer,* 1–11.

47 Jiao, Y., Tran, T.D., and Shi, G. (2021). EffiScene: efficient per-pixel rigidity inference for unsupervised joint learning of optical flow, depth, camera pose and motion segmentation. *IEEE Conference on Computer Vision and Pattern Recognition,* 5538–5547. June 20 2021 to June 25 2021. Nashville, TN, USA.

48 Ranjan, A., Jampani, V., Balles, L. et al. (2019). Competitive collaboration: Joint unsupervised learning of depth, camera motion, optical flow and motion segmentation. *IEEE Conference on Computer Vision and Pattern Recognition,* 12240–12249. June 15 2019 to June 20 2019. Long Beach, CA, USA.

49 Stoffregen, T., Gallego, G., Drummond, T. et al. (2019). Event-based motion segmentation by motion compensation. *IEEE International Conference on Computer Vision,* 7244–7253. Oct. 27 2019 to Nov. 2 2019 Seoul, Korea (South).

50 Chen, D., Li, W., and Hall, P. (2018). Motion estimation and segmentation of natural phenomena. *British Machine Vision Conference,* 1–6. Northumbria University, 3rd–6th September 2018.

51 Sun, D., Sudderth, E.B., and Black, M.J. (2012). Layered segmentation and optical flow estimation over time. *IEEE Computer Society Conference on Computer Vision and Pattern Recognition,* 1768–1775. June 16 2012 to June 21 2012. Providence, RI.

52 Sun, D. (2012). From pixels to layers: Joint motion estimation and segmentation. Doctoral dissertation, Brown University.

53 Chen, X., Member, S., and Pan, L. (2018). A survey of graph cuts/graph search based medical image segmentation. *IEEE Reviews in Biomedical Engineering* 11 (1): 112–124.

54 Khalil, A., Rahman, S.U., Alam, F., and Khalil, I. (2021). Fire detection using multi color space and background modeling. *Fire Technology* 57 (3): 1221–1239.

55 Jarraya, S.K., Hammami, M., Ben-Abdallah, H. (2010). Accurate background modeling for moving object detection in a dynamic scene. *Digital Image Computing: Techniques and Applications,* 52–57.

56 Jiang, B., Meng, H., Zhao, J. et al. (2018). Single image fog and haze removal based on self-adaptive guided image filter and color channel information of sky region. *Multimedia Tools and Applications* 77 (11): 13513–13530.

57 Scharstein, D., Hirschmuller, H., Kitajima, G. et al. (2014). High-resolution stereo datasets with subpixel-accurate ground truth. *German Conference on Pattern Recognition,* 31–42. 2–5 September; Münster, Germany.

58 Butler, D.J., Wulff, J., Stanley, G.B., and Black, M.J. (2012). A naturalistic open source movie for optical flow evaluation. *European Conference on Computer Vision,* 611–625. Florence Italy October 7–13.

59 Sakaridis, C., Dai, D., and Gool, L.V. (2018). Semantic foggy scene understanding with synthetic data. *International Journal of Computer Vision* 126 (9): 973–992.

60 Strabe, K.W. Image sequence server. https://i21www.ira.uka.de/imagesequences/.

61 Wang, Y., Jodoin, P.M., Porikli, F. et al. (2014). Cdnet 2014: an expanded change detection benchmark dataset. *IEEE Computer Society Conference on Computer Vision and Pattern Recognition Workshops,* 387–394. 23 2014 to June 28 2014. Columbus, OH, USA.

3

Robust NLTV-L^1 Optical Flow Estimation Under Adverse Outdoor Conditions

3.1 Introduction

Optical flow estimation and motion segmentation are fundamental challenges in computer vision, forming the basis for various applications including video surveillance and scene understanding [1–3]. Over the past several decades, a wide range of optical flow estimation techniques has been developed [4–7]. Despite these advancements, estimating optical flow in adverse outdoor conditions (e.g., haze, fog, or other atmospheric effects) remains a significant challenge. Likewise, accurately segmenting moving objects in such environments still faces considerable limitations. Both optical flow estimation and motion segmentation are greatly affected by environmental factors, particularly in challenging conditions such as fog, haze, rain, and low-light environments. In these conditions, visibility is greatly reduced, contrast levels are lower, and sudden brightness fluctuations can occur. For instance, fog causes scattering effects that decrease the sharpness of objects' edges, making it difficult for traditional optical flow estimation methods to accurately detect motion boundaries. Rain or haze further complicates the scene by introducing noise that blends moving objects with the background, often resulting in blurred boundaries or inaccurate segmentation of motion fields. Addressing these challenges requires a method that not only accurately captures motion but also effectively preserves edge clarity and resists the smoothing effects that typically occur in such complex visual environments. Generally, most techniques for estimating motion fields fall within the category of variational methods [4, 8–12].

In recent years, significant advancements have been made in variational optical flow estimation methods. Notably, some approaches have replaced L^2 norm with the L^1 norm in total variation (TV)-based optical flow estimation. The TV-L^1 methods [8, 9] generally enforce the flow field in a linear manner, thereby reducing the impact of noise during estimation. However, because TV-L^1 relies solely

Robust Optical Flow Estimation in Outdoor Environments, First Edition. Sana Rao and Rao Kashif.

on local object information, it can lead to excessive smoothing of nonsalient structures within image frames.

To address these limitations, the nonlocal total variational (NLTV-L^1) technique has been proposed for optical flow estimation [4, 13], offering improved robustness in both edge preservation and noise suppression. This enhancement is achieved by incorporating a non-local element into the TV-L^1 model, which serves as an extra smoothness constraint applied to a defined region of the motion field. Incorporating this non-local component allows the algorithm to capture more detailed structural information from objects [14, 15]. A drawback of this non-local strategy is that the pixels within a specified region may not necessarily belong to the same physical surface. To overcome this, [16] introduces a weighting mechanism computed from brightness variations and spatial distances between pixels. Nonetheless, such a weighting scheme remains ineffective against moderate levels of outdoor noise. Consequently, assigning appropriate weights to pixels in a defined neighborhood becomes essential for reliable motion field computation under challenging environmental conditions. Achieving robust moving-object segmentation in complex outdoor scenes is equally demanding due to the presence of blurred motion boundaries.

Robust motion segmentation in complex outdoor environments is further complicated by the blurred motion boundaries of moving objects. There are some works [17–21] solving these problems. Among the variational methods developed to improve optical flow estimation, NLTV-L^1 stands out for its robustness in preserving edges by incorporating nonlocal information. Traditional variational methods rely mostly on local pixel information, which often leads to over-smoothing of edges when handling complex outdoor scenes. In contrast, nonlocal methods, such as NLTV-L^1, consider a broader spatial context within the image, thus capturing relationships between distant pixels. This added context is essential for reducing noise and maintaining sharper motion boundaries, especially in scenes with varying visibility. By expanding beyond local pixel data, NLTV-L^1 offers a solution that better approximates real-world complexities, enabling more accurate flow estimates even when scene details are obscured by environmental interference.

A new robust nonlocal total variational (RNLTV-L^1) flow field estimation method is introduced to overcome the abovementioned challenges. This approach demonstrates robustness in preserving edges, mitigating the issue of over-smoothing near motion boundaries, and effectively segmenting moving objects under complex outdoor conditions, see Figure 3.1. Such capabilities are particularly advantageous for outdoor surveillance systems (especially in environmental changes). The key contributions of this chapter are outlined below: (1) We introduce an enhanced RNLTV-L^1 optical flow estimation technique that incorporates a support weight function designed to prevent blurred motion boundaries

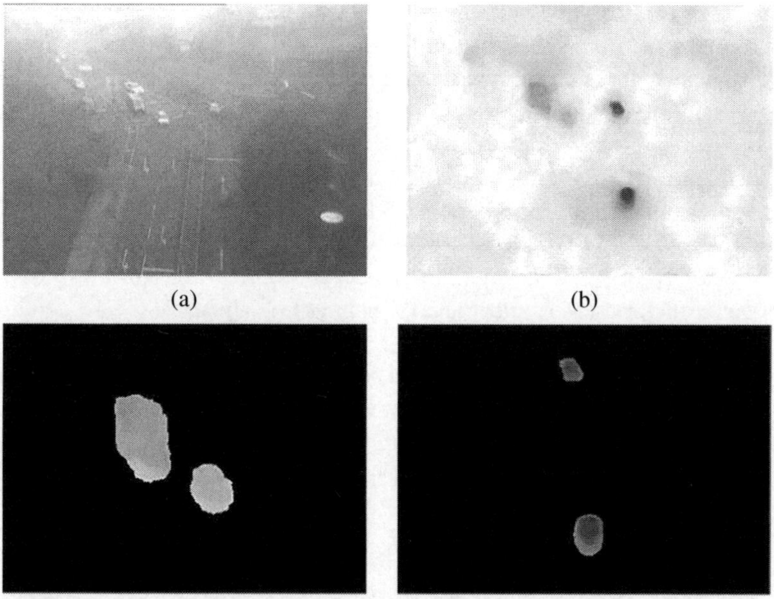

(a) (b) (c) (d)

Figure 3.1 The optical flow estimation and motion segmentation results obtained by the proposed RNLTV-L^1 method in adverse outdoor conditions. (a) The input image frame from the Traffic dataset, (b) the estimated optical flow, and (c, d) the motion pattern-based segmentation of moving objects in different directions (e.g., two-way traffic: there is a difference in the directions of moving vehicles in (c) and (d)). *Source:* Sana Rao (2024) / Springer Nature.

resulting from inappropriate or fixed pixel weights. Additionally, the estimated optical flow is refined through a coarse-to-fine strategy, enhancing its robustness. The effectiveness of the proposed approach is validated not only on two widely used optical flow benchmarks (i.e., the MPI-Sintel and the Middlebury datasets), but also on more challenging real-world datasets, including Foggy Zurich [22], the Traffic [23], and the CDnet2014 [24]. (2) We extend the application of the proposed RNLTV-L^1 method to motion segmentation, enabling the extraction of moving objects from a scene.

3.2 Non-local TV-L^1 for Optical Flow Estimation

3.2.1 Current NLTV-L^1 Optical Flow Estimation Methods

The field of optical flow estimation has a rich history, evolving from basic methods that tracked intensity changes at the pixel level to more advanced variational approaches. Early optical flow methods primarily used local gradient-based techniques that calculated motion based on changes in pixel intensity between frames. While effective for simple scenes, these methods struggled with noise and failed to accurately capture complex motion in real-world conditions. Over time, variational techniques were introduced to address these limitations. By incorporating regularization terms, variational methods could enforce smoothness across the flow field, which reduced noise and stabilized the estimation. The advent of TV-L^1 methods represented a breakthrough, as they applied the L^1 norm to penalize noise more effectively, resulting in sharper motion boundaries. However, even these advanced methods were insufficient for handling adverse outdoor conditions, where environmental factors severely degraded accuracy. The development of NLTV-L^1, which introduces non-local total variation, marked a significant step forward in addressing these limitations, as it allows for improved edge preservation and greater robustness under challenging scenarios.

As discussed in [25], the NLTV-L^1 framework offers strong robustness for optical flow estimation. In the non-local term, the weighting function (given in Eq. (2.7)) can be expressed as:

$$w_{x,y}^{x',y'} = \exp\left\{-\frac{|x-x'|^2 + |y-y'|^2}{2\sigma_S^2} - \frac{|I(x,y) - I(x',y')|^2}{2\sigma_B^2}\right\}. \tag{3.1}$$

Here, $w_{x,y}^{x',y'}$ represents the probability that the neighboring pixel (x', y') and the central pixel (x, y) lie on the same physical surface. This probability is determined using two key factors: the spatial distance σ_S and the brightness distance σ_B between the pixels. In Eq. (3.1), the brightness difference term $|I(x, y) - I(x', y')|^2$ is evaluated alongside the spatial distance term $|x - x'|^2 + |y - y'|^2$ to measure similarity between pixels. On the other hand, the geodesic distance approach requires the neighboring pixel be connected to the central pixel via a path of nearly uniform color. In the case of background pixels, such a path is typically absent, as foreground objects are separated from the background by pronounced edges. To overcome this issue, we introduce a weighting strategy that integrates both the geodesic distance and the pixel color similarity. The strength of using geodesic distance lies in its ability to more effectively link the neighboring pixels to the central pixel, thereby improving optical flow estimation. In the RNLTV-L^1 framework, this support-weight function is tailored to enhance motion boundary precision by adaptively assigning weights to neighboring pixels according to their geodesic closeness and color resemblance. This weighting scheme allows for a more refined understanding of spatial relationships between pixels, which is essential in complex outdoor conditions where visibility is compromised. For instance, when tracking a moving vehicle in foggy conditions, standard distance-based weighting might inadvertently blur the vehicle's boundary due to the low contrast between the vehicle and the foggy background. However, by using geodesic distance, the support-weight function better captures the continuity of the vehicle's structure, preserving its edges and enhancing segmentation accuracy. Additionally, color similarity helps distinguish objects from the background, as it enables the model to detect consistent color patterns even under noisy conditions. Together, these factors make the RNLTV-L^1 approach more robust in preserving details and reducing the likelihood of motion boundary blending, especially in real-world scenes where adverse conditions complicate motion detection.

3.3 RNLTV-L^1 Optical Flow Estimation Method

The process of motion estimation using the RNLTV-L^1 approach, along with the application of the resulting motion field for motion segmentation, can be seen in Figure 3.2.

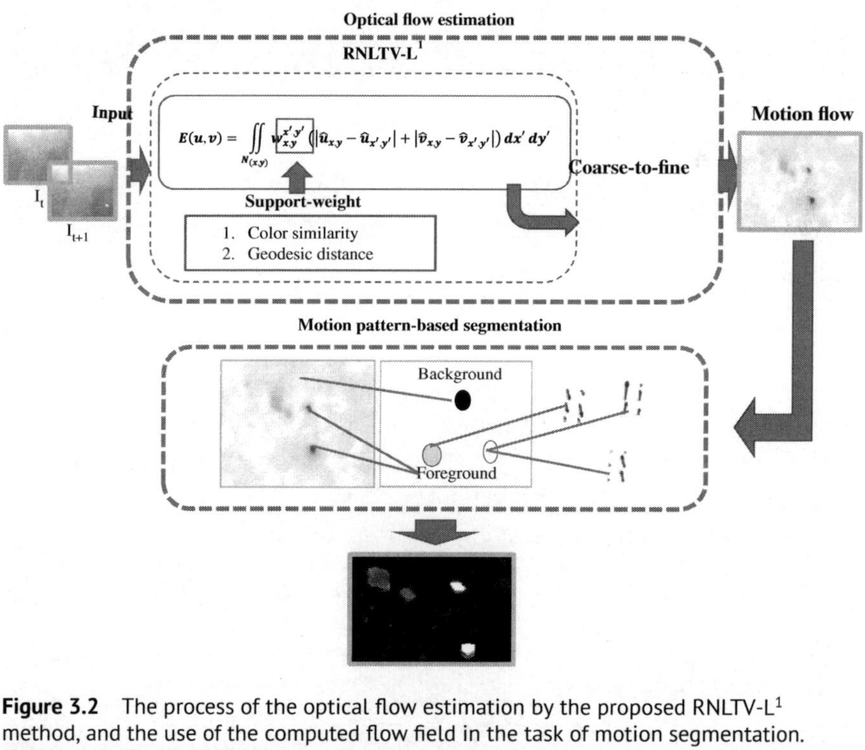

Figure 3.2 The process of the optical flow estimation by the proposed RNLTV-L^1 method, and the use of the computed flow field in the task of motion segmentation. *Source:* Sana Rao (2024) / Springer Nature.

3.3.1 RNLTV-L^1 Optical Flow Estimation Method

To enhance the NLTV-L^1 motion estimation framework, a weighted support function is incorporated to measure the degree of mutual influence among pixels. This function determines pixel weights based on both geodesic distance and the color correspondence. By applying a multi-scale strategy, the weight calculation at each scale effectively minimizes motion ambiguity. The resulting optimization formulation for motion estimation is given as:

$$E(u,v) = \iint_{\Omega} \left\{ \begin{array}{l} \rho(|I_x u + I_y v + I_t| + |\nabla I_x u + \nabla I_y v + \nabla I_t|) + \lambda.\rho(|\nabla u| + |\nabla v|) \\ + \iint_{N_{(x,y)}} w_{x,y}^{x'y'} \left(\left|\hat{u}_{x,y} - \hat{u}_{x'y'}\right| + \left|\hat{v}_{x,y} - \hat{v}_{x'y'}\right| \right) dx'dy' \end{array} \right\} dxdy. \tag{3.2}$$

Here, $|I_x u + I_y v + I_t|$ represents the intensity consistency term, while $|\nabla I_x u + \nabla I_y v + \nabla I_t|$ corresponds to the gradient constancy condition, which introduces additional constraints (i.e., directional motion information) to improve the reliability of the motion estimation process. The operator ∇ denotes the spatial derivative. The Lorentzian cost function, expressed as $\rho(x) = \log\left(1 + \frac{x^2}{2\gamma^2}\right)$, is employed to minimize the impact of outlier values. In Eq. (3.2), $\gamma = 1$ is assigned for the intensity consistency term, $\gamma = 0.03$ for the smoothness constraint, and $\lambda = 0.06$ act as the regularization weight balancing the two terms.

The support-weight function $w_{x,y}^{x'y'}$ in a non-local term can be written as:

$$w_{x,y}^{x'y'} = \exp\left(-\frac{g_{x,y}^{x'y'}}{2\sigma_c^2 n_c}\right), \tag{3.3}$$

where

$$g_{x,y}^{x'y'} = \min_{S \in P_{x,y}^{x'y'}} d(S). \tag{3.4}$$

In Eq. (3.3), the parameter σ_c regulates the impact of color similarity, while n_c represents the count of spectral channels. In Eq. (3.4), $g_{x,y}^{x'y'}$ denotes the minimal route linking the reference pixel to its adjacent pixels, and $P_{x,y}^{x'y'}$ defines the set of all possible routes between these two points. The cost $d(S)$ of a given route is calculated as:

$$d(S) = \sum_{i=2}^{i=n} C((x_i, y_i), (x_{i-1}, y_{i-1})), \tag{3.5}$$

where $C((x_i, y_i), (x_{i-1}, y_{i-1}))$ measures the color similarity between pixels (x, y) and (x', y'), and is defined as:

$$C_{x,y}^{x'y'} = \exp\left(-\frac{|I(x,y) - I(x',y')|^2}{2\sigma_c^2 n_c}\right). \tag{3.6}$$

Figure 3.3 Support-weight computation at three different marked positions (a–c), in the Army sequence of the Middlebury dataset. *Source:* Sana Rao (2024) / Springer Nature.

In Eq. (3.6), chromatic similarity is evaluated between $I(x, y)$ and $I(x', y')$ by using the Euclidean metric in the Lab color domain. Pixels connected through a path exhibiting minimal chromatic variation produce a high affinity weight function, as indicated by the geodesic distance. Figure 3.3 presents an example of edge-aware weighting mechanism, where brighter values along edges correspond to higher weights.

The objective function in Eq. (3.2) consists of two main components: (1) numerical computation of the optical flow and (2) optimization of the computed flow field. The Euler–Lagrange equations for the variables u and v are obtained by minimizing the objective formulation given in Eq. (3.2), and can be expressed as:

$$\begin{cases} \rho'(|I_x u + I_y v + I_t| I_x + |\nabla I_x u + \nabla I_y v + \nabla I_t| \nabla I_x) - \lambda div(\rho'(|\nabla u| + |\nabla v|)|\nabla u|) = 0 \\ \rho'(|I_x u + I_y v + I_t| I_y + |\nabla I_x u + \nabla I_y v + \nabla I_t| \nabla I_y) - \lambda div(\rho'(|\nabla u| + |\nabla v|)|\nabla v|) = 0, \end{cases} \tag{3.7}$$

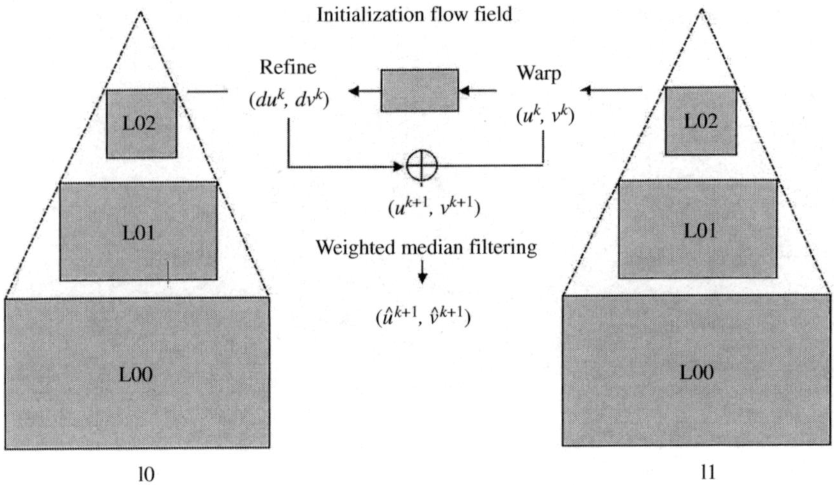

Figure 3.4 The coarse-to-fine process for the optimization of optical flow.

where $div(.)$ represents the optical flow divergence in relation to the spatial coordinates x and y, computed as $div(x,y) = \frac{\partial}{\partial x}u(x,y) + \frac{\partial}{\partial y}v(x,y)$. The set of Euler–Lagrange relations used for estimating motion fields incorporates a nonlinear response term, denoted by $\rho'(.)$. To address this nonlinearity, a multi-scale hierarchical refinement approach is adopted, allowing the motion vectors to be approximated in a more stable, near-linear manner, as shown in Figure 3.4. Let the total number of scales in the image hierarchy be m. The process begins at the lowest-resolution level, where an initial motion estimate is computed and progressively refined as we move toward higher resolutions. At each scale $(1 \le k \le m)$, the initial estimate is represented by $(u^{k^0}, v^{k^0})^T$. For this level, the formulation in Eq. (3.7) is reformulated as:

$$\begin{cases} \rho'^k\left(\left|I_x^k du^k + I_y^k dv^k + I_t^k\right| I_x^k + \left|\nabla I_x^k du^k + \nabla I_y^k dv^k + \nabla I_t^k\right| \nabla I_x^k\right) \\ -\lambda div(\rho'^k \left|\nabla(u^k + du^k)\right|) = 0 \\ \\ \rho'^k\left(\left|I_x^k du^k + I_y^k dv^k + I_t^k\right| I_y^k + \left|\nabla I_x^k du^k + \nabla I_y^k dv^k + \nabla I_t^k\right| \nabla I_y^k\right) \\ -\lambda div(\rho'^k \left|\nabla(v^k + dv^k)\right|) = 0. \end{cases} \tag{3.8}$$

In Eq. (3.8), $(du^k + dv^k)^T$ represents the incremental motion update computed at scale k of the image pyramid. The final motion estimate at this scale becomes $(u^k + du^k,\ v^k + dv^k)$,which serves as the starting point for the next finer scale. In our implementation, the hierarchy consists of 30 scales, and each level is generated using a down-sampling factor $\eta = 2$. The coarsest scale is initialized with $(u^{k^0}, v^{k^0})^T$.

After performing the required inner and outer iterations, the motion increment for the current scale is obtained. The motion field for any given scale is produced by adding the initialization from the previous scale to the newly computed increments, as described in Eq. (3.8). To improve robustness, a weighted median filter is applied to the motion output at each scale to suppress outlier vectors, as illustrated in Figure 3.4. The filtered and refined motion field is then upsampled and used as the initialization for estimating increments at the next scale. This iterative procedure is continued until the convergence criterion is met, as summarized in Algorithm 3.1.

Algorithm 1 A Hierarchical Coarse-to-Fine Scheme Is Adopted for the Proposed RNLTV-L1 Optical Flow Estimation Method

Input: Two consecutive image frames I_1 and I_2, L, u, v, N_{inner}
Output: Optical flow $(\overline{u}, \overline{v})$

Initialize $(u^{k=0}, v^{k=0})$
For $k = 1 \rightarrow L$ do (L = max_level)
  For $w = 1 \rightarrow W$ do ($W = warping$)
    Warp image using (u^{w-1}, v^{w-1})
    Update the weights of pixels by using Eq. (3.3)
    Initialize $(du, dv) \rightarrow 0$
    For $l = 0 \rightarrow N_{inner} - 1$ do
      $(u^{k+1}, v^{k+1}) \leftarrow (u^k, v^k) + (du^{k,l+1}, v^{k,l+1})$
    End For
    Weighted median filtering on (u^{k+1}, v^{k+1})
    $(\overline{u}^{k+1}, \overline{v}^{k+1}) \leftarrow (u^{k+1}, v^{k+1})$
  End For
  Resize $(\overline{u}^{k+1}, \overline{v}^{k+1})$ to the size of the next level
End For
The final flow field $(\overline{u}, \overline{v})$.

In the RNLTV-L[1] method, specific parameters are carefully optimized to achieve a balance between noise reduction and edge preservation. A sensitivity analysis on these parameters showed that lower values of λ enhance smoothness but may lead to overly smoothed boundaries, while higher values reduce noise tolerance. The coarse-to-fine strategy, meanwhile, improves the stability of optical flow estimation by analyzing motion at multiple scales, starting with a low-resolution approximation and progressively refining it at higher resolutions. This strategy is particularly beneficial in scenes with varying scales of motion, as it allows for broader structures to be captured early on, with finer details added at later stages. By mitigating motion ambiguity across levels, this multilevel approach makes the

RNLTV-L^1 method more resilient to sudden environmental changes and provides a smoother, more reliable motion flow estimation.

3.4 Motion Pattern-based Segmentation

Motion pattern-based segmentation is an important technique for isolating moving objects in a scene based on their motion characteristics. Unlike traditional methods that rely solely on pixel intensity or texture differences, motion pattern-based segmentation takes advantage of optical flow fields, making it particularly effective for dynamic environments. Motion pattern-based segmentation is essential in various real-world applications where dynamic scenes contain multiple objects moving independently, such as in traffic monitoring or sports event analysis. Unlike static scene analysis, which relies on stationary cues, dynamic environments require algorithms that can adapt to shifting patterns and isolate individual objects based on their distinct motion characteristics. This segmentation technique effectively manages variations in speed, trajectory, and object deformation, making it invaluable for tracking and analyzing complex, fast-moving scenes. The focus on optical flow-based patterns provides an adaptive approach that captures subtle motion which might be missed by methods relying solely on static pixel intensity or texture.

3.4.1 Introduction to Motion Pattern Segmentation

Motion pattern segmentation works by clustering pixels with similar motion vectors into distinct regions, representing individual moving objects or parts of the scene. The segmentation process analyzes the optical flow fields, identifying coherent areas where motion vectors show consistent patterns, like translation, rotation, or deformation. This clustering of motion vectors allows the method to adaptively identify and group objects within a scene, even as their relative motion varies over time. For example, in scenes with multiple moving vehicles, the segmentation can separate cars traveling at different speeds and directions, assigning each a distinct region based on motion consistency. The approach is highly versatile, as it can also distinguish more complex patterns, such as rotating objects or groups of people moving with a unified direction in crowd analysis. By focusing on motion characteristics, this method effectively handles variations in lighting or background noise, which can complicate segmentation when relying on appearance-based features alone.

Motion pattern segmentation plays a crucial role in object tracking, scene understanding, and video surveillance. By segmenting different moving objects, these methods enable efficient motion analysis, contributing to various real-world applications such as autonomous driving, crowd monitoring, and sports analytics. In particular, segmenting motion patterns becomes critical when dealing with complex scenes, such as multiple objects moving at varying speeds or in different directions.

3.4.2 Methods for Identifying Motion Patterns

Several methods can be employed to identify motion patterns in optical flow data. Traditional approaches involve clustering techniques, where optical flow vectors are grouped based on their similarities. Methods such as K-means clustering, mean-shift, or graph-based approaches are often used to group pixels according to the direction and magnitude of their motion vectors. Clustering techniques like K-means, for instance, are useful for grouping pixels with uniform motion vectors, but may struggle with complex, high-noise environments. Mean-shift clustering can improve on this by detecting high-density regions, providing a more flexible segmentation in scenarios where motion is highly variable. Graph-based approaches add another layer of robustness by considering spatial relationships between pixels, offering improved performance in scenes where object boundaries are ambiguous or overlapped. These clustering techniques are limited, however, in their ability to adapt to scenes with significant noise or complex motion, which has led to the integration of deep learning techniques for enhanced accuracy.

Alternatively, variational methods can be used to directly minimize the energy of motion discontinuities. Variational methods often incorporate regularization terms that promote smooth motion flow, reducing the impact of random noise. By optimizing these motion discontinuities, these methods can clearly outline the boundaries between objects and background areas in the scene. These optimization approaches excel in structured scenes where motion boundaries are clearly defined but can struggle in highly dynamic environments with overlapping motions. Despite their strengths in traditional scenarios, variational methods are being increasingly supplemented by deep learning approaches, which offer enhanced adaptability to irregular motion patterns. These methods often involve formulating an optimization problem to find the motion boundaries between objects.

In recent years, deep learning techniques have proven highly effective for motion pattern segmentation. Architectures such as fully convolutional networks (FCNs) and region-based convolutional neural networks (RCNNs) can be trained on large datasets to automatically identify moving objects by learning motion-specific features. These neural network models often outperform traditional techniques, especially when dealing with noisy or low-quality optical flow fields.

Layer-based motion segmentation methods [17–21] often struggle to accurately delineate the structure and boundaries of moving objects in challenging outdoor environments. This limitation is largely due to the difficulty of extracting an optimal foreground under varying illumination conditions. Illumination changes pose significant challenges in dynamic outdoor environments, as moving objects can blend with the background due to shadow, glare, or low-light conditions. In such cases, traditional layer-based methods struggle to define clear motion boundaries, often leading to inaccurate segmentation results. The proposed RNLTV-L^1 method overcomes this limitation by incorporating a support-weight function that assigns weights based on color similarity and geodesic distance. This weighting scheme enables the method to adjust dynamically to varying illumination, preserving object boundary details even under challenging conditions. The integration of non-local information also helps distinguish subtle motion cues, enabling accurate segmentation of moving objects even when illumination varies across the scene. To alleviate this problem, the RNLTV-L^1 method is employed for motion segmentation, dividing the scene into foreground (FL) and background layers (BL). Owing to its robustness in adverse outdoor environments, the method effectively captures the directional motion of objects, representing it as the FL, while zero motion vectors correspond to the BL. This relationship can be formulated as:

$$M_f = \alpha_f FL_f + (1 - \alpha_f) BL_f, \tag{3.9}$$

Here, $\alpha_f \in [0, 1]$ denotes the opacity value of a flow field. This formulation effectively groups pixels exhibiting similar motion directions, enabling the separation of moving objects from the background. Within this layered framework, the motion flow field functions as a robust FL representation, highlighting the movement of objects against a static or gradually varying background. The assignment of null vectors to BLs reduces computational complexity and isolates moving objects more accurately. This layered segmentation, defined mathematically in Eq. (3.9), efficiently groups objects moving in similar directions, even in complex outdoor environments where background noise is significant. This approach is particularly beneficial for applications such as urban traffic analysis, where vehicles moving in different directions and speeds need to be accurately segmented from the background.

3.5 Incorporating Deep Learning for Motion Segmentation

Deep learning has revolutionized motion segmentation by leveraging its capability to learn intricate patterns from extensive datasets. Convolutional neural networks

(CNNs) and their variants, such as U-Nets or segmentation networks, have been particularly successful at performing pixel-wise motion segmentation. Deep learning methods transform motion segmentation by leveraging large datasets to learn complex, multi-dimensional motion patterns that are challenging to define manually. These models can automatically detect relevant features, adapting to different environments and improving over time with more data. For instance, CNN-based architectures can learn spatial patterns effectively, while U-Nets add precision to boundary detection through their skip connections, which retain high-resolution details. Recurrent models, such as recurrent neural networks (RNNs) and long short-term memories (LSTMs), introduce the advantage of temporal context, ensuring that each frame in a sequence builds on previous motion patterns, leading to smoother, more accurate segmentation over time.

These models are trained on labeled datasets, where the ground truth includes both motion vectors and segmentation maps. The network learns to associate certain motion characteristics with distinct object boundaries. Through extensive training, these models develop a deep understanding of how motion patterns correlate with object characteristics, such as shape and size, improving segmentation fidelity even in noisy or low-contrast environments. Advanced architectures can also leverage multi-scale processing, where motion features are analyzed at different resolutions, providing a comprehensive view of both large-scale motion and finer details. This multi-scale capability helps deep learning models generalize better to unfamiliar scenes, making them highly adaptable for applications like autonomous navigation and surveillance, where unpredictable motion patterns are common. Additionally, RNNs and LSTM architectures can be employed to maintain temporal consistency, enabling motion patterns to evolve smoothly across frames.

Deep learning models outperform traditional methods in cases where the motion patterns are complex, or the scene contains multiple overlapping motions. They can also generalize better to new environments by learning robust motion features from the training data.

3.6 Fusion of Appearance and Motion-based Features

In many cases, combining both appearance-based features (such as texture and color) with motion-based features from optical flow results in superior segmentation performance. These multi-modal approaches exploit the complementary strengths of both methods, with motion features helping to segment moving objects and appearance features refining the boundaries.

For example, in scenes with weak or ambiguous motion cues (like when objects are temporarily stationary), appearance-based features provide additional information to improve segmentation. Feature fusion techniques such as concatenation or attention mechanisms can be used to combine both types of features within deep learning architectures. Feature fusion techniques like attention mechanisms enable the model to selectively focus on the most relevant cues, dynamically weighting appearance and motion data to enhance segmentation accuracy. In a multi-modal deep learning framework, motion patterns help identify moving regions, while appearance features refine the contours of objects. For example, when a vehicle slows down at an intersection, its motion cues weaken, potentially leading to misclassification. By incorporating appearance features, the model retains the vehicle's outline based on its color and shape, ensuring consistency in segmentation. This fusion of modalities not only enhances the model's robustness but also enables it to handle a wider range of scenarios, where motion alone might not provide sufficient segmentation cues. The fusion of appearance and motion data is particularly useful in applications like action recognition, object tracking, and video segmentation, where motion patterns alone may not be sufficient to segment complex objects.

3.7 Experiments and Analysis

In this section, we assess the performance of the RNLTV-L^1 optical flow estimation method several challenging datasets that feature complexities such as intricate motions and adverse outdoor conditions.

3.7.1 The Influence of Parameters

This section investigates how the choice of free parameters influences the performance of the RNLTV-L^1 method. In particular, we assess the impact of λ (i.e., the weighting factor for the smoothing term) and η (i.e., the down-sampling ratio) on the accuracy of the flow field, as illustrated in Figures 3.5–3.7. It is observed that the value of λ significantly influences (e.g., smaller values generally lead to more precise flow field estimations). The marked areas in Figure 3.5 show the smoothness around the object boundaries. Based on the results obtained by our proposed method, we recommend to set $\lambda = 2$ for the RubberWhale and Alley-1 image sequences of the Middlebury and the MPI-Sintel datasets. Moreover, to examine the influence of various values of λ on the performance of our proposed RNLTV-L^1 method, we summarize the Average Angular Error (AAE) and End-Point Error (EPE) in Figure 3.7.

We further examine how varying the down-sampling factor and adjusting the number of warping iterations at each pyramid level affect the results. To assess

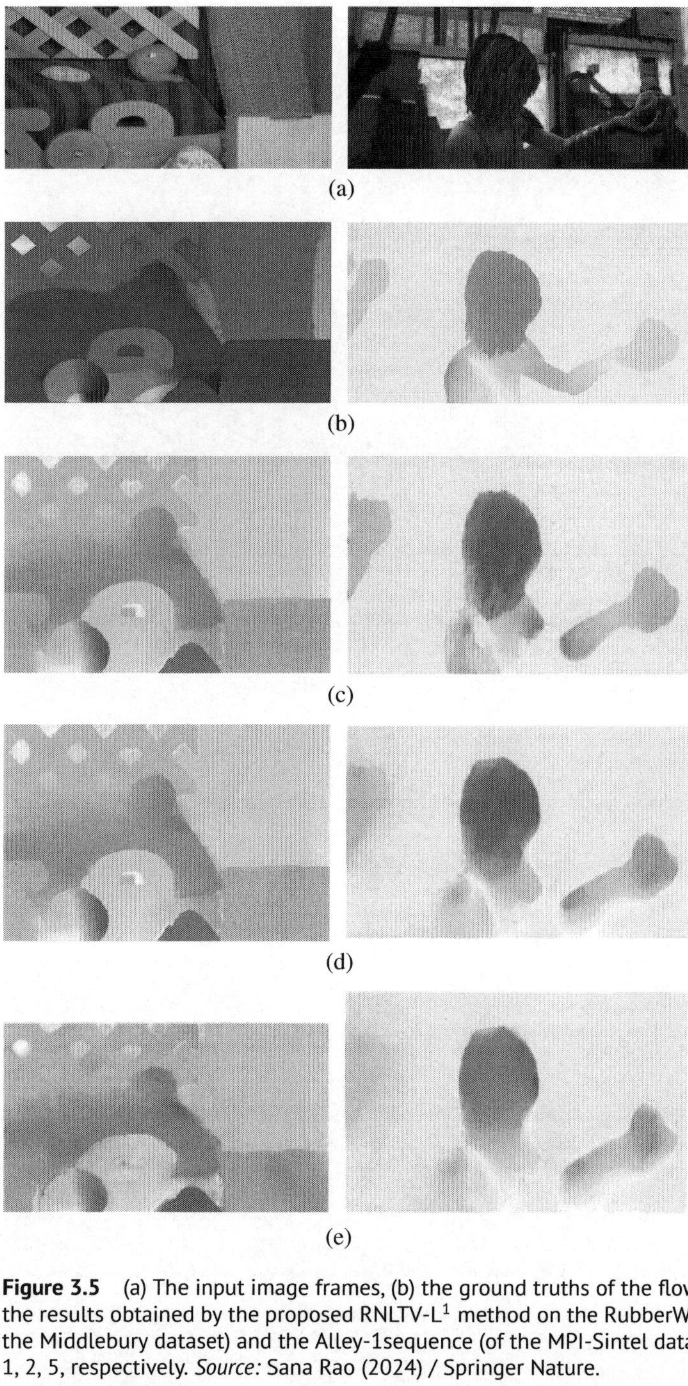

Figure 3.5 (a) The input image frames, (b) the ground truths of the flow fields, and (c–e) the results obtained by the proposed RNLTV-L^1 method on the RubberWhale sequence (of the Middlebury dataset) and the Alley-1sequence (of the MPI-Sintel dataset) with $\lambda = 0.5$, 1, 2, 5, respectively. *Source:* Sana Rao (2024) / Springer Nature.

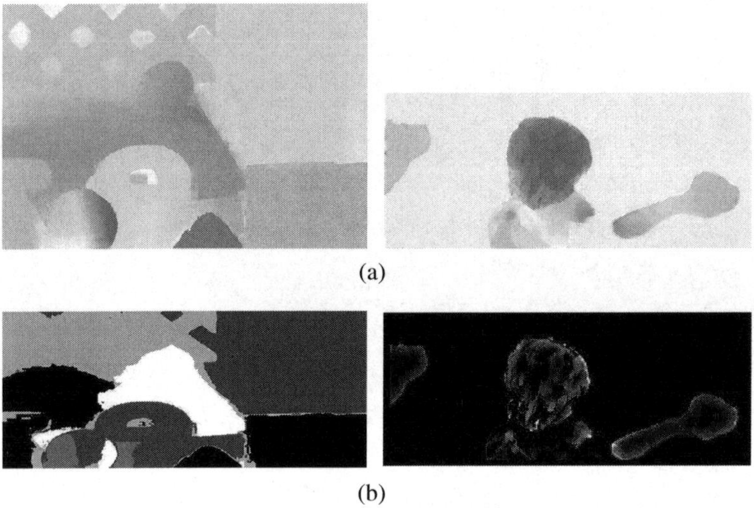

Figure 3.6 The optical flow estimation and the motion pattern-based segmentation results obtained by the proposed RNLTV-L^1 method with $\lambda = 2$ and the down-sampling factor η =2 on the RubberWhale sequence (of the Middlebury dataset) and the Alley-1 sequence (of the MPI-Sintel dataset). (a) The estimated flow field, and (b) the motion pattern-based segmentation, respectively. *Source:* Sana Rao (2024) / Springer Nature.

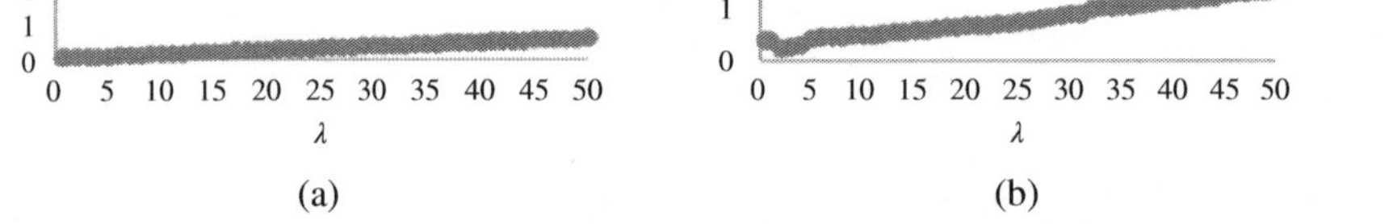

Figure 3.7 The AAE and EPE obtained by the proposed RNLTV-L^1 method on (a) the RubberWhale sequence (of the Middlebury dataset) and (b) the Alley-1 sequence (of the MPI-Sintel dataset) by using various values of λ in the RNLTV-L^1 method.

the impact of varying down-sampling factors on the performance, the AAE and EPE metrics are presented in Figure 3.8. A comparison between the RubberWhale and Alley-1 sequences reveals that the RubberWhale sequence involves only small displacement motions, making it better suited to a lower down-sampling factor.

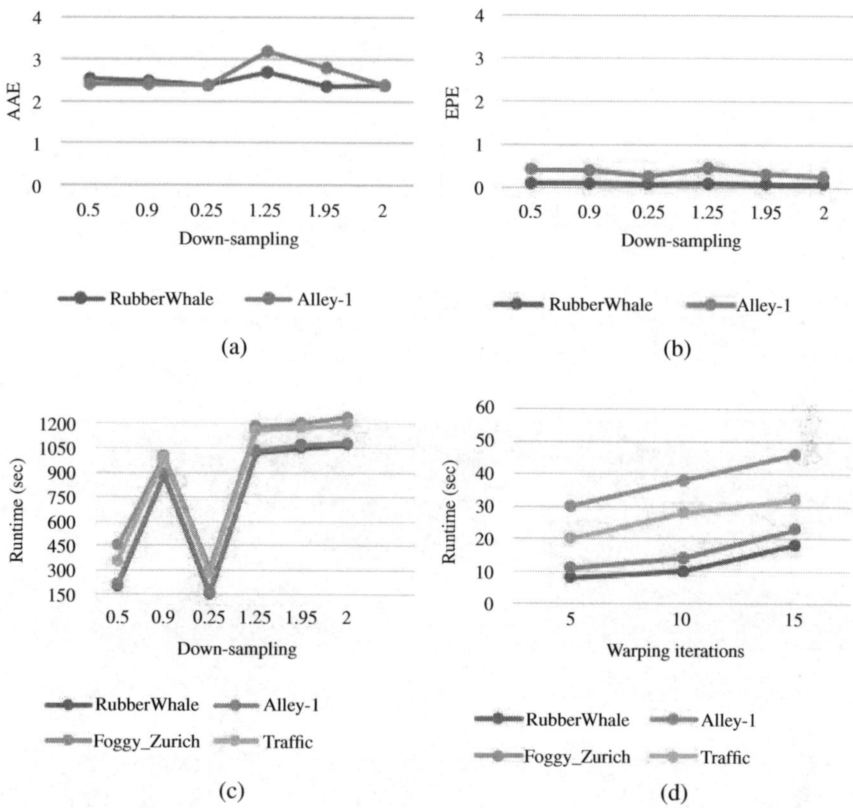

Figure 3.8 (a–d) The AAE, EPE, and run time analysis of the proposed RNLTV-L^1 optical flow estimation method with respect to various down-sampling factor values and number of warping iterations.

3.8 Results

To thoroughly evaluate the robustness of RNLTV-L^1, the method was tested on datasets that represent a wide range of environmental challenges, such as MPI-Sintel (which features animated sequences with varying motion complexity), Middlebury (with high-contrast indoor and outdoor scenes), Foggy Zurich (highlighting the effects of dense fog), Traffic (representing urban environments with complex moving objects), and CDnet2014 (focusing on surveillance scenarios with various weather conditions). These datasets were specifically chosen to cover adverse conditions that are common in real-world applications, including different levels of fog density, low-light settings, and challenging backgrounds. Using standard evaluation metrics like EPE and angular error, the method's ability to handle adverse conditions was objectively quantified, providing a comprehensive view of its performance under diverse environmental scenarios.

3.8.1 Results on the Middlebury Dataset

Experiments are conducted on the Middlebury dataset to evaluate the effectiveness of the proposed RNLTV-L^1 method in comparison with existing approaches. The method is benchmarked against several state-of-the-art optical flow estimation approaches, including PMC [25], JOF [26], AGFOF [27], NNF+EAC [28], TVG [29], RNLOD [4], and LDOF [6]. The qualitative and quantitative outcomes for the Middlebury dataset are presented in Figure 3.9 and Table 3.1.

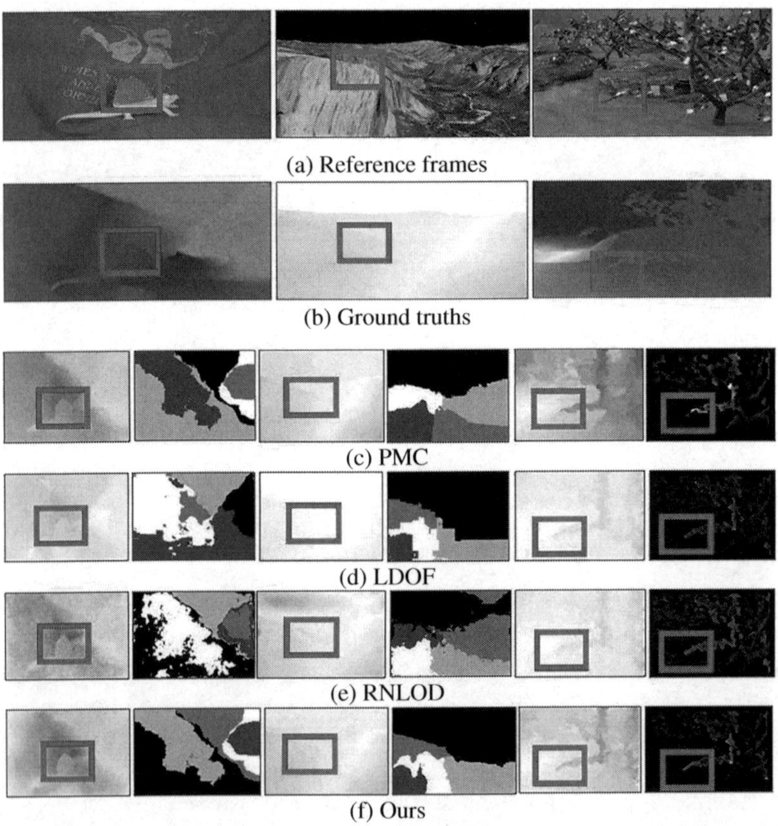

Figure 3.9 (a–f) Visual comparison of the optical flow estimation and motion segmentation results obtained by the PMC, LDOF, RNLOD methods, and the proposed RNLTV-L^1 method on three complex image sequences of the Middlebury dataset, where the grey squares indicate the boundary problems. *Source:* Sana Rao (2024) / Springer Nature.

Table 3.1 The table reports the performance of eight competing optical flow estimation methods across six challenging image sequences from the Middlebury dataset, with bold values indicating the lowest AAE and EPE errors.

	Dimetrodon		Yosemite		Grove2		Grove3		Venus		RubberWhale	
Method	**AAE**	**EPE**	**AAE**	**EPE**	**AAE**	**EPE**	**AAE**	**EPE**	**AAE**	**EPE**	**AAE**	**EPE**
PMC (SPIC2022) [25]	2.56	**0.13**	2.45	0.14	2.83	0.18	4.98	0.41	4.14	0.26	2.69	0.09
JOF (TMM2020) [26]	**2.37**	0.15	3.12	0.19	2.70	0.15	6.18	0.68	5.10	0.37	2.33	0.07
AGFOF (SP 2016) [27]	2.56	0.13	2.87	0.16	2.80	0.16	8.19	0.89	7.10	0.49	3.72	0.13
LDOF (TPAMI2010) [6]	3.38	0.20	2.46	0.12	3.52	0.20	**4.47**	**0.37**	6.20	0.40	3.80	0.14
NNF+EAC (PR2017) [28]	2.64	0.13	2.36	0.14	2.70	0.19	5.06	0.48	4.96	0.29	4.26	0.12
RNLOD (TIP2017) [4]	3.45	0.35	3.00	0.16	2.68	0.15	6.07	0.62	4.38	0.27	2.39	0.07
TVG (ICBD2018) [29]	2.92	0.30	2.32	0.13	2.79	0.19	7.09	0.69	5.28	0.30	4.60	0.15
RNLTV-L^1	2.80	0.37	**2.25**	0.13	**2.68**	0.15	4.98	0.41	**4.09**	**0.22**	**2.28**	**0.06**

Figure 3.9 visualizes the results obtained by the proposed RNLTV-L^1 method on several challenging image sequences. This approach demonstrates strong robustness in preserving edges, mitigating the issue of excessive smoothing near motion boundaries, and accurately segmenting moving objects. As shown in Figure 3.9, it delivers the best performance on the Yosemite and Grove image sequences, owing to its effective preservation of edges and motion boundaries.

As indicated in Table 3.1, the RNLTV-L^1 method achieves lower AAE and EPE values for several image sequences, with the exception of the Dimetrodon sequence, which involves complex nonrigid motion in textureless regions. For the Yosemite and Grove sequences, the method delivers competitive AAE and EPE performance. Table 3.1 summarizes key methods in optical flow estimation, comparing their performance under adverse conditions. While traditional TV-L^1 methods are effective in noise reduction, they often lack the necessary edge-preserving capabilities in complex scenes with low contrast. Standard NLTV-L^1 methods improve upon this by incorporating broader spatial context, allowing for better preservation of motion boundaries. However, these methods still struggle under highly variable visibility conditions, where motion boundaries tend to blur or blend with the background. In comparison, RNLTV-L^1 integrates a support weight function that assigns adaptive weights to pixels based on geodesic distance and color similarity, significantly improving its ability to maintain sharp motion boundaries even in foggy, rainy, or hazy scenarios. By introducing this support-weight function, RNLTV-L^1 offers a more comprehensive solution that effective manages the challenges posed by diverse outdoor environments.

Table 3.2 The results obtained by the six competing methods on five complex image sequences of the MPI-Sintel dataset. The bold values represent the lowest AAE and EPE errors.

	Alley-1		Bamboo		Mountain		Market		Cave-2	
Method	**AAE**	**EPE**	**AAE**	**EPE**	**AAE**	**EPE**	**AAE**	**EPE**	**AAE**	**EPE**
LDOF (TPAMI2010)	3.30	0.40	4.80	0.39	12.50	1.83	18.43	2.85	7.61	0.50
NNF+EAC (PR2017)	2.18	0.20	4.27	0.23	12.50	1.89	**8.72**	**1.48**	**5.03**	1.53
TVG (ICBD2018)	2.44	0.19	4.25	0.21	20.55	2.02	14.31	2.89	5.28	**0.30**
RNLOD (TIP2017)	2.99	0.37	4.39	0.27	17.77	2.34	25.97	6.38	17.26	3.43
Classic+nl (CVPR2010) [30]	**2.06**	**0.18**	5.17	0.36	9.36	**1.39**	12.96	2.34	15.38	3.27
RNLTV-L^1	2.39	0.25	**4.17**	**0.20**	**8.80**	1.48	12.90	2.47	5.18	1.73

3.8.2 Results on the MPI-Sintel Dataset

We then evaluate the RNLTV-L^1 method on the MPI-Sintel dataset, which is more challenging due to its illumination changes and complex outdoor scenes. In this evaluation, the method is compared with several state-of-the-art methods, including, LDOF, NNF+EAC, TVG, RNLOD, and Classic+nl-fast.

Table 3.2 presents the AAE and EPE values for several image sequences from the MPI-Sintel dataset. The RNLTV-L^1 method achieves lower AAE and EPE for the Bamboo and Mountain sequences, demonstrating its robustness in optical flow estimation. However, our proposed method produces high AAE and EPE results on the Market and Cave-2 image sequences, possibly because of the low-contrast challenge in the image sequences.

Figure 3.10 presents the reference image frames, corresponding ground-truths and flow field results produced by six optical flow estimation methods for two complex image sequences from the MPI-Sintel dataset. It results show that the RNLTV-L^1 method delivers robust performance on the Alley-1 and Bamboo image sequences. The Bamboo sequence contains too much edge detail, hence they may not be estimated accurately by the LDOF method, as can be seen in the marked areas in Figure 3.10. However, the proposed RNLTV-L^1 method and the Classic+nl-fast method can robustly preserve the edges near the motion boundaries.

3.8.3 Results on the Complex Outdoor Datasets

The RNLTV-L^1 method is evaluated on more challenging outdoor datasets and compared with other recent methods. Figures 3.11 and 3.12 show the obtained

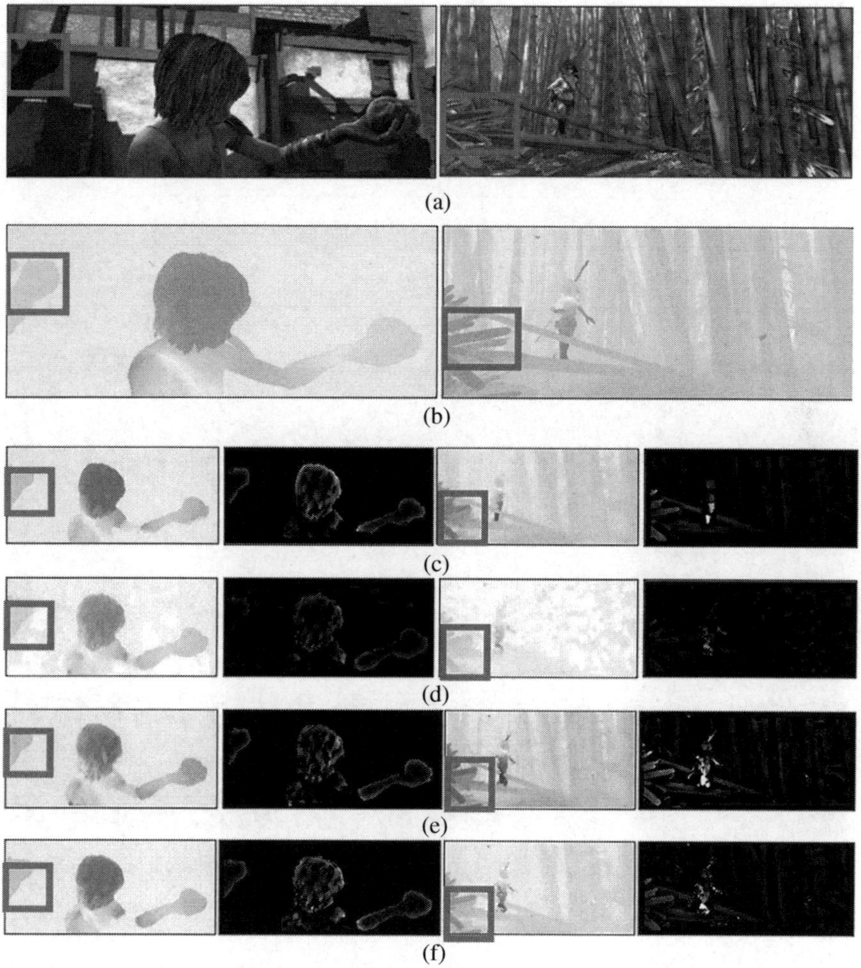

Figure 3.10 Visual evaluation of the flow field and motion segmentation results obtained by four competing methods on the Alley-1 and Bamboo sequences of the MPI-Sintel dataset. (a) The input image frames. (b) The ground truths of optical flow, (c–f) the optical flow estimation (left) and segmentation (right) results obtained by the Classic+nl-fast, LDOF, RNLOD, and the proposed RNLTV-L^1 methods, respectively. *Source:* Sana Rao (2024) / Springer Nature.

qualitative results. Unfortunately, most outdoor datasets lack ground-truth data for challenging scenarios. To address this limitation, the results is presented to illustrate the impact of adverse outdoor conditions on moving objects. As shown in Figures 3.11 and 3.12, the flow field estimates generated by the proposed RNLTV-L^1 method exhibit higher accuracy than those produced by several

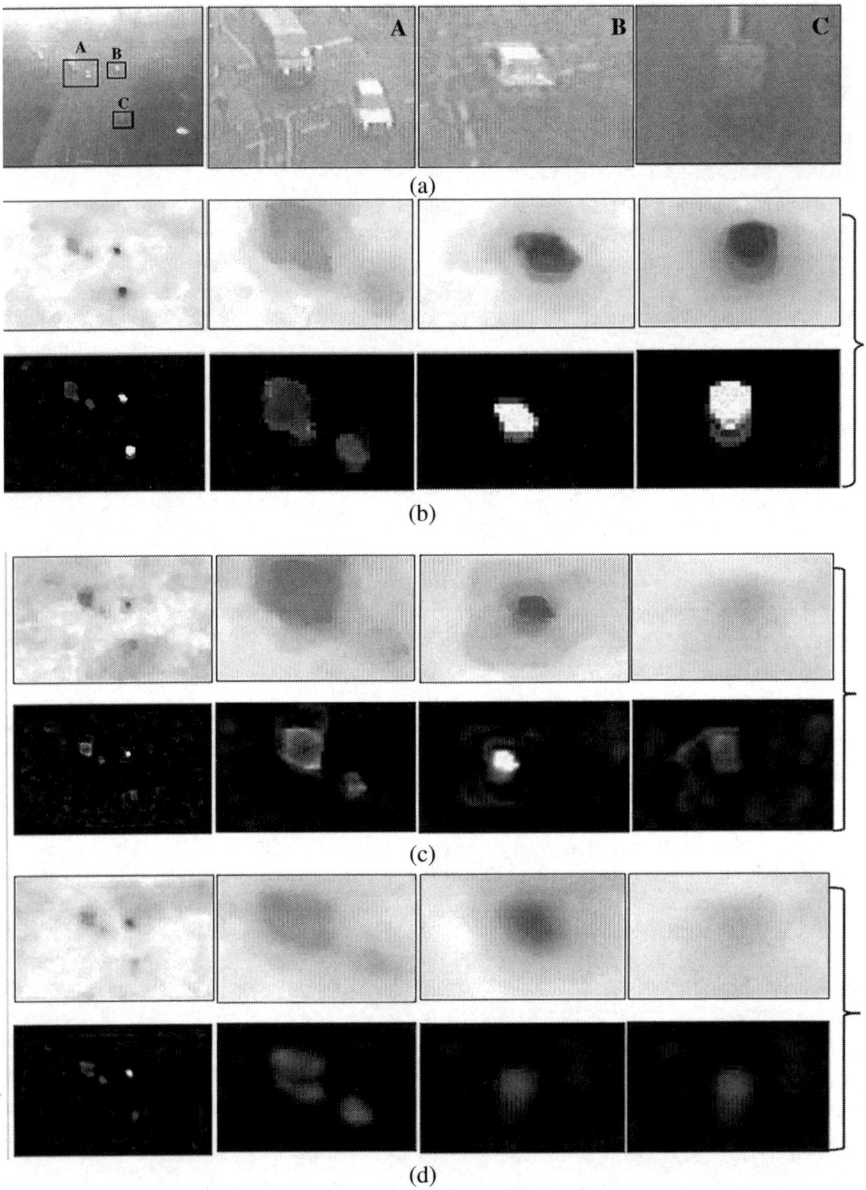

Figure 3.11 Visual comparison of the flow field and motion segmentation results obtained by three competing methods. (a) The input image sequence of the Traffic dataset with an amplified view of the marked moving objects. (b–d) The optical flow estimation (top) and segmentation (bottom) results obtained by the proposed method, the Classic+nl-fast, and the RNLOD methods, respectively. *Source:* Sana Rao (2024) / Springer Nature.

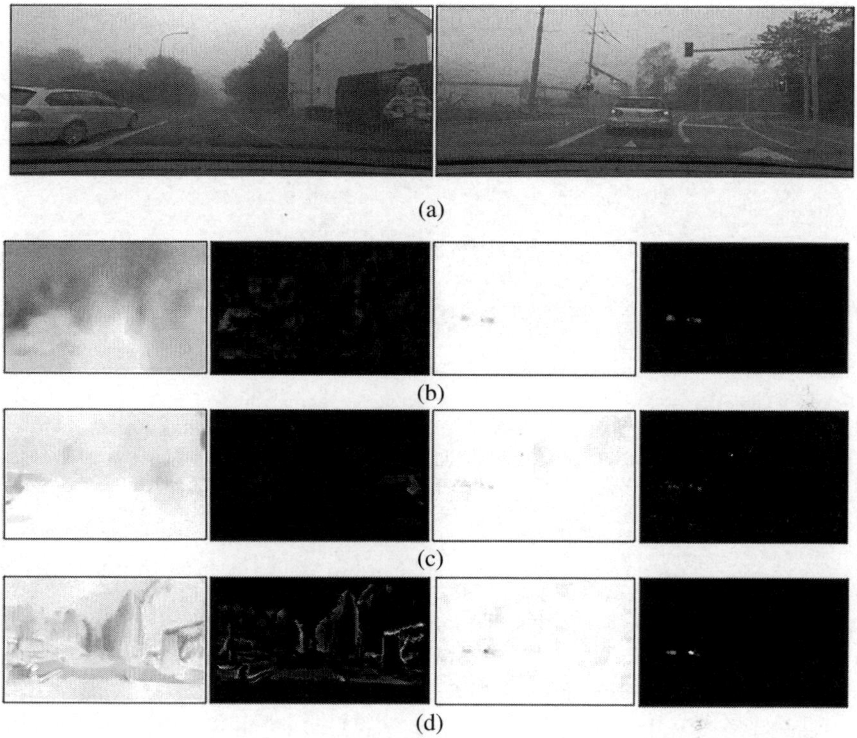

Figure 3.12 Visual evaluation of the obtained flow fields and corresponding motion segmentation results from three competing techniques. (a) The input image sequences of the Foggy Zurich dataset and (b–d) the optical flow estimation (left) and segmentation (right) results obtained by the Classic+nl-fast, LDOF, and the proposed RNLTV-L^1 methods, respectively. *Source:* Sana Rao (2024) / Springer Nature.

competing approaches. The image sequences involve complex outdoor environments (e.g., haze and fog). The enlarged view of the moving car in Figure 3.11 demonstrate that the support-weight function in the proposed method effectively alleviates over-smoothness, particularly near the edges and boundaries of moving objects in adverse conditions.

3.8.4 Results on the CDnet2014 Dataset

The CDnet2014 dataset has more challenging outdoor scenarios. When evaluated against several state-of-the-art methods, the RNLTV-L^1 optical flow estimation method achieves the best overall results. Figure 3.13 illustrates optical flow estimation and motion segmentation for a moving car in the highway sequence under varying weather conditions. While the SOF method estimates the optical

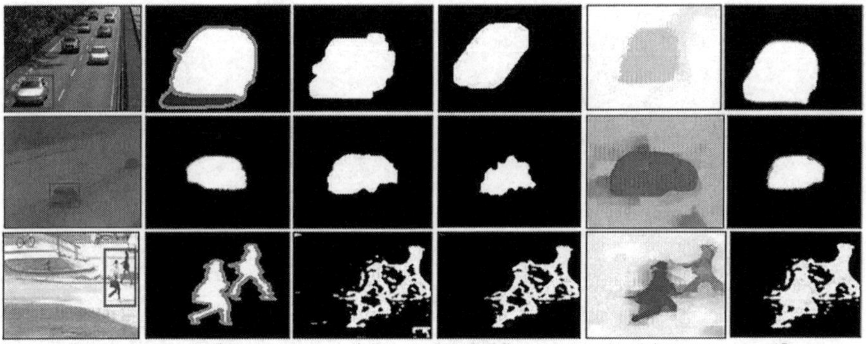

Figure 3.13 The optical flow estimation and motion segmentation results for the CDNet2014 dataset. (a) The input image sequences with marked moving objects. (b) The ground truths of segmented moving objects. (c–d) The motion segmentation results by the SOF, and TPF methods, respectively. (e, f) The optical flow estimation (left) and segmentation (right) results obtained by the proposed RNLTV-L^1 method. *Source:* Sana Rao (2024) / Springer Nature.

flow accurately, it tends to merge certain moving foreground objects with the background in such conditions. The sharp edges and the motion boundaries of moving objects can be easily observed from the optical flow estimation and motion segmentation results. Although minor errors are present around the legs of the pedestrians in Figure 3.13, the shape and position of the pedestrian can be easily recognized from the results obtained by our proposed method.

The enhanced robustness of the RNLTV-L^1 method makes it highly suitable for real-world applications where environmental conditions are unpredictable. In autonomous driving, for instance, accurate optical flow estimation is critical for detecting and tracking objects on the road, particularly in adverse weather. By providing reliable motion segmentation even in fog or rain, RNLTV-L^1 can contribute to safer navigation, reducing the likelihood of accidents due to unclear object boundaries. In surveillance, the method's ability to accurately segment moving objects in low-light or foggy environments aids in reliable monitoring, enhancing security in both urban and rural settings. Environmental monitoring is another potential application, as RNLTV-L^1 can be used to track changes in natural scenes over time, helping detect objects or animals in challenging weather, which is essential for studying and preserving ecosystems.

3.9 Conclusion

This chapter introduces the RNLTV-L^1 method, developed to preserve the edges of motion regions in moving objects. The estimated flow field is employed to segment target objects to their motion patterns, even under adverse outdoor conditions. The method is evaluated on two widely used optical flow estimation benchmarks (i.e., the MPI-Sintel and the Middlebury datasets) as well as three real-world outdoor datasets (i.e., the Foggy Zurich, the Traffic, and the CDnet2014). Across all evaluations, the approach consistently achieves the highest accuracy and overall performance when compared to several competing methods, with particularly strong results on challenging outdoor datasets. Experimental findings also demonstrate that the RNLTV-L^1 method notably improves the accuracy of motion segmentation.

Nonetheless, the RNLTV-L^1 method can still produce errors in scenarios where fine edge details are crucial for accurate optical flow estimation under outdoor conditions. This limitation arises because the local filtering step implemented using a weighted median filter may not be sufficiently robust to maintain motion boundaries in such environments. To address this, the following chapter introduces an enhanced NLTV-L^1optical flow estimation method incorporating a weighted guided filter (NLTV-L^1+WGF). This filter, built on a linear model, offers both computationally efficient and strong edge-preserving capabilities in complex outdoor scenarios (i.e., haze, fog, etc.).

The RNLTV-L^1 method could be adapted to include dynamic weighting schemes that adjust in real time based on detected environmental conditions. For example, a machine learning model could be integrated to analyze scene properties and adjust support-weight parameters automatically, optimizing segmentation for various weather scenarios. Additionally, this method's robustness in adverse conditions is likely to inspire further research in adaptive optical flow estimation techniques, encouraging developments that consider a wider range of environmental factors. By addressing real-world challenges, RNLTV-L^1 sets a precedent for future work in computer vision, emphasizing the need for models that can reliably perform in diverse and unpredictable environments.

Activities

Activity 1: Frame Annotation

- Provide two consecutive video frames taken during rainy conditions.
- Annotate areas where motion vectors might be most accurate using NLTV-L^1.

Activity 2: Error Analysis

- Give optical flow results obtained with L^2 norm and L^1 norm under foggy conditions.
- Analyze and compare errors, noting regions where L^1 norm performs better.

Activity 3: Simulation

- Provide sample optical flow data with added Gaussian noise.
- Ask students to clean the data using NLTV regularization principles and compare results.

Activity 4: Group Debate

Topic: Is NLTV-L^1 the ultimate solution for optical flow under adverse outdoor conditions, or are hybrid models better?

Exercises

Fill in the Blanks

1) NLTV in optical flow estimation stands for ____________
2) The L^1 norm is preferred in optical flow estimation because it is more robust to____________
3) Adverse outdoor conditions like fog and rain introduce ____________ that NLTV-L^1 estimation must handle.
4) The ____________ term in the optical flow equation ensures temporal smoothness between frames.
5) A significant challenge in optical flow under low-light conditions is ____________ detection.

Answers:

1) Nonlocal total variation
2) Noise and outliers
3) Noise and distortions
4) Regularization
5) Motion or edge

Multiple Choice (MCQs)

1 What is the main advantage of using NLTV in optical flow estimation?

a. Faster computation
b. Better
c. Simplifies the model
d. Reduces contrast

Answer. b

2 Which norm is more sensitive to outliers in optical flow estimation?
  a. L^1
  b. L^2
  c. L-infinity
  d. L^0
  Answer. b

3 In NLTV-L^1 estimation, regularization helps by:
  a. Removing all motion
  b. Adding random noise
  c. Suppressing distortion
  d. Increasing frame rate
  Answer. c

4 What type of outdoor condition is NLTV-L^1 particularly robust against?
  a. High contrast
  b. Perfect lighting
  c. Rain and fog
  d. Indoor lighting
  Answer. c

5 Which component of NLTV-L^1 optical flow estimation helps handle variations in pixel intensity?
  a. Regularization
  b. Data fidelity
  c. Motion vectors
  d. Gradient smoothing
  Answer. a

Short Questions

Q1 Explain why optical flow estimation is more challenging under adverse outdoor conditions.
Answer. Noise, low visibility, and unpredictable distortions make it harder to accurately detect motion.

Q2 How does the L^1 norm reduce errors caused by noise?
Answer. By minimizing absolute differences, the L^1 norm prevents outliers from dominating the error calculation.

Q3 What is the primary function of regularization in optical flow?
Answer. To ensure smoothness and consistency in motion estimation across adjacent pixels.

Q4 Identify one real-world application of NLTV-L^1 optical flow estimation.
Answer. Autonomous vehicle navigation in bad weather.

Q5 What are the two main components of the NLTV-L^1 optical flow framework?
Answer. NLTV regularization and the L^1 data fidelity term.

Long Questions and Scenarios

Q1 Discuss the steps involved in NLTV-L^1 optical flow estimation and explain how it handles noise in rainy conditions.
Answer. Include frame alignment, gradient calculation, NLTV regularization, and the role of the L^1 norm in minimizing outlier impact.

Q2 Create a diagram showing motion vectors for an object moving in a foggy environment. Label how NLTV and L^1 norms refine the flow.
Answer. Students should draw vectors with noise at edges, showing smoother, corrected vectors after NLTV-L^1 processing.

Q3 Compare NLTV-L^1 optical flow estimation with a deep learning-based optical flow method. Highlight pros and cons.
Answer. NLTV-L^1 offers robustness and simplicity, while deep learning methods require training and high computational resources but can handle complex scenarios.

Q4 Develop a hypothetical scenario where NLTV-L^1 fails. Suggest improvements or hybrid techniques to address the failure.
Answer. Scenario: Heavy motion blur combined with snowflakes. Improvements: Integrate temporal smoothing and use auxiliary sensor data.

Q5 Solve the optical flow equation using given gradients and explain the role of each term in the NLTV-L^1 framework.
Answer. Provide sample derivatives (I_x, I_y, I_t) and walk through solving u and v while explaining the mathematical role of regularization and fidelity terms.

Examples with Solutions

Given the nonlocal weight function in Eq. (3.1):

$$w_{x,y}^{(x'y')} = \exp\left(-\frac{(|x-x'|^2+|y-y'|^2)}{2\sigma_S^2} - \frac{|I(x,y)-I(x',y')|^2}{2\sigma_B^2}\right),$$

(a) Show that when $\sigma_S \rightarrow \infty$, the weight depends only on pixel intensity similarity.
(b) Show that when $\sigma_B \rightarrow \infty$, the weight depends only on spatial distance.

Solution:

(a) If $\sigma_S \rightarrow \infty$, the spatial term denominator $\rightarrow \infty$, so $\exp(-0)$. The weight reduces to:

$$w = \exp\left(-\frac{|I(x,y)-I(x',y')|^2}{2\sigma_S^2}\right).$$

So, only spatial proximity matters.
Thus, RNLTV-L^1 balances both intensity and spatial proximity simultaneously.

Applied – Motion Segmentation

In motion pattern-based segmentation, optical flow vectors of two cars are detected:

Car A motion vector: $(u_A, v_A) = (5, 0)$.

Car B motion vector: $(u_B, v_B) = (0, -4)$.

Use K-means clustering ($k = 2$) on these motion vectors. Show why these two cars are assigned to distinct clusters.

Solution:

- Distance between vectors:

$$d = \sqrt{(u_A-u_B)^2+(v_A-v_B)^2} = \sqrt{(5-0)^2+(0-(-4))^2} = \sqrt{25+16} = \sqrt{41}.$$

- Since the distance is large, K-means separates them into different clusters → Car A (moving horizontally), Car B (moving vertically).
- This demonstrates motion pattern segmentation: separating objects based on distinct optical flow directions.

Energy Function Analysis

Consider the RNLTV-L1 energy function, why is the Lorentzian cost function

$$\rho(x)\log\left(1+\frac{x^2}{2\gamma^2}\right)$$

preferred over a quadratic penalty in outdoor noise conditions?

Solution:

- Quadratic penalty (x^2) heavily penalizes outliers (e.g., sudden brightness spikes due to glare).
- Lorentzian grows slower for large x making it **robust to outliers** → does not let noise dominate optimization.
- Hence, it stabilizes optical flow estimation in **noisy outdoor scenes** (fog, rain).

Algorithm Understanding

Refer to Algorithm 3.1 (coarse-to-fine scheme). Why is a hierarchical multi-scale strategy better than directly estimating optical flow at full resolution?

Solution:

- At full resolution, large displacements cause mismatches.
- Multi-scale strategy:
  1) Estimate coarse motion at low resolution.
  2) Refine with finer details progressively.
- This prevents local minima, reduces computational cost, and stabilizes estimation in complex scenes.
- Especially useful in traffic datasets (large motion differences across vehicles).

Segmentation Model – Practical

Equation defines motion segmentation:

$$M_f = \alpha_f FL_f + (1-\alpha_f)BL_f$$

If $\alpha f = 0.8$, what does this imply about the segmentation result?

Solution:

- $\alpha f = 0.8$ → foreground contributes 80%, background contributes 20%.
- Means strong confidence that pixels belong to moving objects.
- High αf → clear separation of moving vehicles/pedestrians.
- Low αf → more background influence (less motion confidence).

Computing Nonlocal Weights in Foggy Conditions

Problem:

Two pixels $p_1$and p_3 in a foggy road scene must be compared for similarity in the RNLTV-L^1 model. How are their weights computed?

Step 1: Define Feature Vectors

Each pixel has intensity value I, and spatial coordinates (x, y).

- p_1: $I = 120, (x = 15, y = 40)$
- p_2: $I = 128, (x = 18, y = 41)$

Step 2: Compute Intensity Difference

$$d_I = |I_{p1} - I_{P2}| = |120 - 128| = 8$$

Step 3: Compute Spatial/Geodesic Distance

$$d_s = \sqrt{(18 - 15)^2 + (41 - 40)^2} = \sqrt{9 + 1} = \sqrt{10} \approx 3.16$$

Step 4: Apply Gaussian Weighting

$$w(p1, p2) = \exp\left(-\frac{d_I^2}{\sigma_I^2}\right) . \exp\left(-\frac{d_s^2}{\sigma_s^2}\right)$$

Assume $\sigma_i = 10$, $\sigma_s = 5$.

- Intensity term: $\exp(-64/100) \approx 0.53$
- Spatial term: $\exp(-10/25) \approx 0.67$

Final weight: $0.53 \times 0.67 = \mathbf{0.35}$

Conclusion:

Even under fog, nonlocal weights capture similarity. Lower value means p_2 contributes moderately to p_1's flow estimation.

(Insert small diagram: two foggy pixels with arrows, weight label = 0.35)

Edge Preservation versus TV-L^1

Scenario:

- A car edge in haze, pixels on either side have intensities 90 and 150.
- TV-L^1 tends to average across the edge → blurred boundary.
- RNLTV-L^1 computes weights and preserves discontinuity.

Step 1: TV-L^1 Gradient Cost

$$Cost_{TV} = |150 - 90| = 60$$

Step 2: RNLTV-L^1 Weighted Cost

Edge weight decays sharply across discontinuity. Assume $w = 0.1$.

$$Cost_{RNLTV} = w.|150 - 90| = 0.1.60 = 6$$

Result:

TV-L^1: blurred edges (higher penalty).
RNLTV-L^1: preserves motion boundary (low penalty).

Outlier Suppression with Lorentzian Penalty

Problem:

Optical flow outlier due to rain streak, measured flow error = 15 pixels.

Step 1: L^2 Norm Cost

$$E_{L2} = 15^2 = 225$$

Step 2: L^1 Norm Cost

$$E_{L1} = |15| = 15$$

Step 3: Lorentzian Penalty (ρ)

$$\rho(e) = \log 1 + \frac{e^2}{2\sigma^2}$$

Take $\sigma = 5$

$$\rho(e) = \log\left(1 + \frac{225}{50}\right) = \log 1 + 4.5 = \log 5.5 \approx 1.70$$

Coarse-to-Fine Flow Refinement

Scenario:

Two frames:

- Vehicle moves 2 px/frame (slow) in background.
- Pedestrian moves 8 px/frame (fast) in foreground.

Step 1: Low Resolution (1/4 Scale)

- Vehicle visible (2 px $\rightarrow$ 0.5 px).
- Pedestrian almost lost (8 px $\rightarrow$ 2 px).

Step 2: Next Resolution (1/2 Scale):

- Vehicle refined to 1 px.
- Pedestrian detected at 4 px.

Step 3: Full Resolution

- Vehicle = 2 px.
- Pedestrian = 8 px.

Motion Segmentation from Flow

Problem:
Scene has 3 moving objects:

- Car flow magnitude ≈ 5 px
- Pedestrian ≈ 2 px
- Background trees ≈ 0.3 px (wind sway)

Step 1: Collect Flow Vectors

$F = \{5.1, 4.8, 5.2, 2.1, 1.9, 0.3, 0.4\}$

Step 2: Apply *k*-means Clustering (*k* = 3)

- **Cluster 1:** Car (≈5 px)
- **Cluster 2:** Pedestrian (≈2 px)
- **Cluster 3:** Background (≈0.3 px)

Step 3: Segment Objects
Label pixels based on cluster membership.

Flow Estimation Under Varying Illumination

Problem:
Two consecutive frames:

- **Frame 1:** Pixel intensity = 100
- **Frame 2:** Pixel intensity = 120

But the object did not move; brightness increased due to sunlight.

Step 1: Brightness Constancy Assumption (BCA)

$$I(x, y, t) = I(x + u, y + v, t + 1)$$

Fails here because $100 \neq 120$.

Step 2: Robust NLTV-L^1 Photometric Invariance
Instead of raw intensity, use gradient constancy:

- **Gradient at Frame 1:** $\partial I/\partial x = 15$
- **Gradient at Frame 2:** $\partial I/\partial x = 16$

Handling Occlusion in Pedestrian Crossing

Problem:
Two pedestrians cross paths:

- P_1 flow = 3 px right
- P_2 flow = 3 px left

At overlap region, which flow should model choose?

Step 1: Data Term Ambiguity
Both flows possible → high error.

Step 2: Nonlocal Regularization
Weights computed with neighborhoods:

- Pixels matching P_1 → weight 0.7
- Pixels matching P_2 → weight 0.3

Step 3: Final Estimate
Weighted decision favors P_1's flow (rightward) at overlap pixel.

Robustness to Rain Streaks

Scenario:
Optical flow between two frames of street camera in rain. Rain streaks move fast downward ($\approx$ 10 px), background cars move slowly ($\approx$ 2 px).

Step 1: Raw Flow Computation
Rain dominates → average motion $\approx$ 6 px downward (wrong).

Step 2: Robust L^1 Penalty with NL Weighting

- Car flow neighborhood similarity strong (weights $\approx$ 0.9).
- Rain streak similarity weak (weights $\approx$ 0.1).

Step 3: Corrected Flow
Weighted average $= 0.9 \cdot 2 + 0.1 \cdot 10 = 2.8 \approx$ car motion.

Flow Field Smoothing with Edge-Aware Weighting

Problem:
Two adjacent regions in foggy traffic scene:

- Region A (car body): smooth motion 5 px
- Region B (road background): 0 px motion

Step 1: Without Edge-Preservation
Smoothing across boundary → flow values become 2–3 px → blurry.

Step 2: With Edge-Aware NL Weights
Across strong intensity edge, weight ≈ 0.05 (suppresses smoothing).
Within car region, weight ≈ 0.9 (strong smoothing).

Step 3: Final Flow
Car = 5 px preserved; Road = 0 px preserved.

Computational Complexity Reduction

Scenario:
Image = 512 × 512 pixels = 262,144 computations.
Direct pairwise NL weights → $O(N^2) \approx 68$ billion ops (too high).

Step 1: Approximation via Patch-based Neighborhoods
Use patch radius = 5 → compare only 11 × 11 = 121 neighbors per pixel.

Step 2: New Complexity
$\approx N \times 121 \approx 31$ million ops (manageable).

References

1 Wu, M., Li, X., Liu, C. et al. (2019). Robust global motion estimation for video security based on improved k-means clustering. *Journal of Ambient Intelligence and Humanized Computing* 10 (2): 439–448.
2 Lejmi, W., Mahjoub, M.A., and Ben, A.K. (2017). Event detection in video sequences: challenges and Perspectives. *International Conference on Natural Computation, Fuzzy Systems and Knowledge Discovery*, 29 July 2017 - 31 July 2017 Guilin University of Electronic Technology. 682–690.
3 Inkawhich, N., Inkawhich, M., Chen, Y., and Li, H. (2018). Adversarial attacks for optical flow-based action recognition classifiers, arXiv preprint: 1811.11875.
4 Zhang, C., Chen, Z., Wang, M. et al. (2017). Robust non-local tv-l1 optical flow estimation with occlusion detection. *IEEE Transactions on Image Processing* 26 (8): 4055–4067.
5 Mei, L., Lai, J., Xie, X. et al. (2019). Illumination-invariance optical flow estimation using weighted regularization transform. *IEEE Transactions on Circuits and Systems for Video Technology* 30 (2): 495–508.
6 Sánchez, P.J., Meinhardt-Llopis, E., and Facciolo, G. (2013). Tv-l1 optical flow estimation. *Image Process Line* 3 (1): 137–150.

7 Brox, T. and Malik, J. (2010). Large displacement optical flow: descriptor matching in variational motion estimation. *IEEE Transactions on Pattern Analysis and Machine Intelligence* 33 (3): 500–513.

8 Werlberger, M., Pock, T., and Bischof, H. (2010). Motion estimation with non-local total variation regularization. *IEEE Computer Society Conference on Computer Vision and Pattern Recognition,* June 13 2010 to June 18 2010. San Francisco, CA. 2464–2471.

9 Belhachmi, Z. and Hecht, F. (2016). An adaptive approach for the segmentation and the tv-filtering in the optic flow estimation. *Journal of Mathematical Imaging and Vision* 54 (3): 358–377.

10 Bengtsson, T., McKelvey, T., and Lindström, K. (2017). On robust optical flow estimation on image sequences with differently exposed frames using primal–dual optimization. *Image and Vision Computing* 57 (1): 78–88.

11 Yang, J. and Li, H. (2015). Dense, accurate optical flow estimation with piecewise parametric model. *IEEE Computer Society Conference on Computer Vision and Pattern Recognition*, June 7 2015 to June 12 2015 Boston, MA, USA. 1019–1027.

12 Doshi, H. and Kiran, N.U. (2021). On an edge-preserving variational model for optical flow estimation, arXiv preprint: 2207.10302.

13 Said, A.B., Hadjidj, R., and Foufou, S. (2019). Total variation for image denoising based on a novel smart edge detector: an application to medical images. *Journal of Mathematical Imaging and Vision* 61 (1): 106–121.

14 Li, Y. and Osher, S. (2009). A new median formula with applications to pde based denoising. *Communications in Mathematical Sciences* 7 (3): 741–753.

15 Yoon, K.J. and Kweon, I.S. (2006). Adaptive support-weight approach for correspondence search. *IEEE Transactions on Pattern Analysis and Machine Intelligence* 28 (4): 650–656.

16 Chen, J., Cai, Z., Lai, J., and Xie, X. (2018). A filtering-based framework for optical flow estimation. *IEEE Transactions on Circuits and Systems for Video Technology* 29 (5): 1350–1364.

17 Anthwal, S. and Ganotra, D. (2019). An overview of optical flow-based approaches for motion segmentation. *The Imaging Science Journal* 67 (5): 284–294.

18 Soumya, T. (2008). A moving object segmentation method for low Illumination night videos. *World Congress on Engineering and Computer Science,* San Francisco, USA, from October 22–24, 2008. 1161–1165.

19 Chen, D., Li, W., and Hall, P. (2018). Motion estimation and segmentation of natural phenomena. *British Machine Vision Conference,* Newcastle, UK, September 3–6, 2018. 1–6.

20 Jarraya, S.K., Hammami, M., Ben-Abdallah, H. (2010). Accurate background modeling for moving object detection in a dynamic scene. *Digital Image Computing: Techniques and Applications,* 52–57.

21 Unger, M., Werlberger, M., Pock, T., and Bischof, H. (2012). Joint motion estimation and segmentation of complex scenes with label costs and occlusion modeling. *IEEE Computer Society Conference on Computer Vision and Pattern Recognition,* June 16 2012 to June 21 2012. Providence, RI. 1878–1885.

22 Sakaridis, C., Dai, D., and Gool, L.V. (2018). Semantic foggy scene understanding with synthetic data[J]. *International Journal of Computer Vision* 126 (9): 973–992.

23 Strabe, K.W. Image sequence server. https://i21www.ira.uka.de/imagesequences/.

24 Wang, Y., Jodoin, P.M., Porikli, F. et al. (2014). Cdnet 2014: an expanded change detection benchmark dataset. *IEEE Computer Society Conference on Computer Vision and Pattern Recognition Workshops,* June 23 2014 to June 28 2014. Columbus, OH, USA. 387–394.

25 Zhang, C., Feng, C., Chen, Z. et al. (2022). Parallel multiscale context-based edge-preserving optical flow estimation with occlusion detection. *Signal Processing: Image Communication* 101 (1): 116545–116560.

26 Tu, Z., Poppe, R., and Veltkamp, R.C. (2016). Adaptive guided image filter for warping in variational optical flow computation. *Signal Processing* 127 (1): 253–265.

27 Tu, Z., Xie, W., Cao, J. et al. (2017). Variational method for joint optical flow estimation and edge-aware image restoration. *Pattern Recognition* 65 (1): 11–25.

28 Zhang, Y., Fan, H., Zheng, J., and Zhang, C. (2018). Improving the optical flow accuracy based on the total variation of local-global method. *IEEE International Conference on Big Data, Big Data,* July 2–7, 2018, San Francisco, CA, USA. 4658–4664.

29 Zhang, C., Ge, L., Chen, Z. et al. (2019). Refined tv-l1 optical flow estimation using joint filtering. *IEEE Transactions on Multimedia* 22 (2): 349–364.

30 Sevilla-Lara, L., Sun, D., Jampani, V., and Black, M.J. (2016). Optical flow with semantic segmentation and localized layers. *IEEE Computer Society Conference on Computer Vision and Pattern Recognition,* June 27 2016 to June 30 2016 Las Vegas, NV, USA. 3889–3898.

4

Robust Optical Flow Estimation via Edge-preserving Filter

4.1 Introduction

The optical flow estimation of moving objects from image sequences is a classical task in computer vision. During the past few decades, significant progress has been achieved in this area [1–5]. However, accurate and robust optical flow estimation in complex outdoor conditions remains a challenging problem. Early optical flow estimation methods [4, 6] are derived from the variational formulation based on two basic assumptions (i.e., the brightness constancy assumption and the smoothness assumption). However, these methods are sensitive to adverse outdoor conditions.

To overcome the abovementioned problems, the nonlinear filters and the local filters (including the median filter [MF], the weighted median filter [WMF], the bilateral filter [BF], and the guided filter [GF]) are used as a pre-processing step, or they are used with a nonlocal term in the variational optical flow estimation methods to smooth the flow field or to remove the outliers [2, 3, 7–9]. The local filters are generally simple. However, the local filters may not preserve the sharp edges of moving objects, and thus they may produce halo artifacts. One major reason leading to the halo artifacts is the local property and the invariable regularization parameters of the local filters. To address this problem, Ref. [7] presents an edge-aware weighting function to adjust the regularization parameters and incorporate the function in the GF adaptively.

The local and nonlinear filters used in variational optical flow estimation methods have significantly improved the accuracy of flow field calculation. In some state-of-the-art optical flow estimation methods, the median filtering of flow fields during optimization effectively alleviates the influence of outliers and makes the estimated flow boundaries more robust [10–13]. However, the MF and the WMF optimization may cause the problem of edge blurring. Recently, the robust non-local total variational (RNLTV-L^1) method [14] has been proposed that is more

Robust Optical Flow Estimation in Outdoor Environments, First Edition. Sana Rao and Rao Kashif.

robust in terms of edge preservation and noise reduction. Specifically, a nonlocal term with a total variation (TV)-L^1 is introduced as an additional smoothness constraint to perceive additional information of an object. However, the local filters may produce the problem of halo artifacts around the motion boundaries

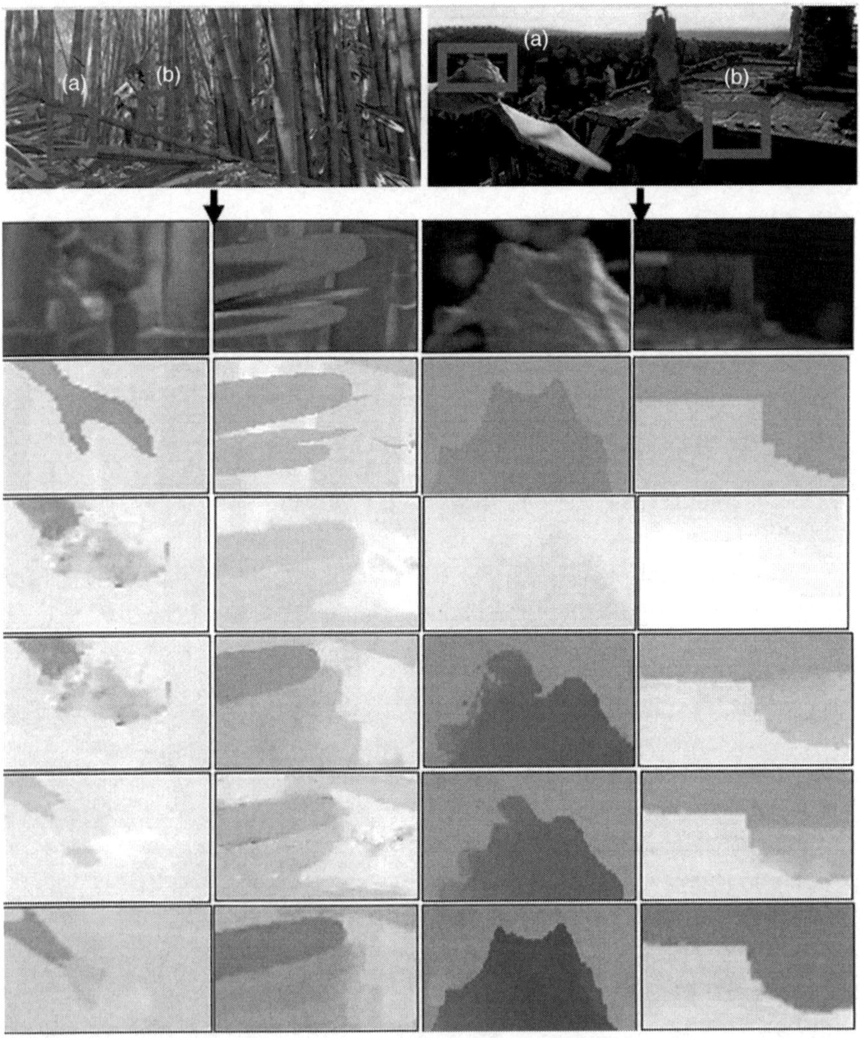

Figure 4.1 Visualization of limitations of the MF, WMF, and GF filters for the variational optical flow estimation methods. (a) The reference image frames (RF) of the MPI-Sintel dataset. (b) The ground-truths (GT) of the optical flow and the results obtained by the MF, WMF, GF, and WGF, respectively. *Source:* Sana Rao et al., (2021) / with permission of Elsevier.

[7, 9, 15, 16]. The challenge of the GF optimization for optical flow estimation [3, 8] is the structure inconsistency between the source and target image frames. Figure 4.1 shows some results obtained by the variational optical flow estimation methods with the aforementioned filtering techniques under outdoor complex conditions. The results show that the motion boundaries are blurred compared to the ground-truths.

To address the abovementioned problems, we propose an edge-preserving non-local total variation (NLTV)-L^1+ weighted guided filter (WGF) method for optical flow estimation. We introduce a robust WGF [7] to NLTV-L^1 to optimize the optical flow using the coarse-to-fine strategy.

4.2 Robust WGF for the Optical Flow Estimation

In this section, we introduce the robust WGF designed to preserve edges in optical flow estimation. The core idea behind the WGF is inspired by human visual perception: larger weights are assigned to pixels located at edges, as opposed to those in flat areas of an image. This ensures that critical details, such as motion boundaries, are maintained throughout the filtering process.

4.2.1 Robust WGF

The key assumption of the robust WGF is based on human visual perception. Larger weights are assigned to pixels at the edges of an image than to pixels in the flat areas of the image. The local variance of the pixels in a 3×3 window in a guidance image is applied to compute the weight of each pixel. The local variance of a pixel is normalized as in [7]. A guidance image I is used, which could be identical to the image F to be filtered. The output image F_k is calculated by implementing a linear transformation of a guidance image I_k in a window ω_i centered at the pixel i:

$$F_k = a_i I_k + b_i \quad \forall k \in \omega_i, \tag{4.1}$$

where a_i and b_i are assumed to be constants in ω_i. The values of a_i and b_i are then obtained by minimizing a cost function $E(a_i, b_i)$, which can be written as:

$$E(a_i, b_i) = \sum_{k \in \omega_i} \left[(a_i I_k + b_i - X_k)^2 + \frac{\rho}{\psi_I(i)} a_i^2 \right], \tag{4.2}$$

where

$$\psi_I(i) = \frac{1}{N} \sum_{k=1}^{N} \frac{\sigma_{I,1}^2(i) + \lambda}{\sigma_{I,1}^2(k) + \lambda}. \tag{4.3}$$

In Eq. (4.2), the function consists of two terms: the data term $(a_i I_k + b_i - X_k)^2$ and the smoothness term ρa_i^2. The function can adaptively adjust the regularization parameter by employing the edge-aware weight function $\psi_I(i)$. In Eq. (4.3), the symbol λ is a small positive constant, and its value is selected as $(0.001 \times L)^2$, where L is the dynamic range of the input image. I is a guidance image with N pixels, and $\sigma_{I,1}^2(i)$ and $\sigma_{I,1}^2(k)$ are the variances of I in the 3×3 windows centered at pixels i and k, respectively. All pixels in the guidance image I are used in the computation of ψ_I. Moreover, the edge-aware weight function $\psi_I(i)$ measures the edge strength of a pixel concerning the whole guidance image I. The value is usually larger than 1.0 for a pixel at an edge of an image, and it is smaller than 1.0 when a pixel is located in a flat area. The solution is obtained by minimizing the cost function $E(a_i, b_i)$ defined in Eq. (4.2). The optimal values of a_i and b_i are computed as:

$$a_i = \frac{\frac{1}{N}\sum_{k\in\omega_i} I_k X_k - \overline{I}_i \overline{X}_i}{\sigma_i^2 + \frac{\rho}{\psi_I(i)}} \tag{4.4}$$

$$b_i = \overline{X}_i - a_i \overline{I}_i, \tag{4.5}$$

where σ_i^2 indicates the variance of the guidance image I in ω_i, and $\overline{X}_i$ and $\overline{I}_i$ are the means of the input image X and the guided image I, respectively. With the estimated linear coefficients a_i and b_i, the output image F_k can be computed as:

$$F_k = \overline{a}_k I_k + \overline{b}_k, \tag{4.6}$$

where $\overline{a}_k = \frac{1}{|\omega_k|}\sum_{i\in\omega_k} a_i$, and $\overline{b}_k = \frac{1}{|\omega_k|}\sum_{i\in\omega_k} b_i$ are the mean values of a_i and b_i in ω_k.

4.2.2 Parameter Influence and Optimization

We evaluated the robust WGF on the image sequences of the MPI-Sintel dataset to show the influence of parameters (i.e., the window radius r and the smoothing parameter ρ). The output image obtained by the robust WGF using different parameter values indicates that suitable parameter values can help to preserve the edges of objects, while improper values may lead to blurring at boundaries. To achieve accurate edges, the robust WGF requires a suitable window radius r and the smoothing parameter ρ.

To evaluate the influence of different parameters, the window radius r is set as 2, 4, or 8, and the parameter ρ is set as 0.1^2, 0.2^2 and 0.4^2, as shown in Figure 4.2. It can be noted that the value of ρ plays an essential role in the generation of halo artifacts and noise. The noise becomes more visible in the results obtained by the robust WGF when the value of ρ is increased. This indicates that the robustness of the WGF increases with a decrease in ρ value. We compare the GF with the WGF, and employ Peak Signal-to-Noise Ratio (PSNR) to evaluate their objective

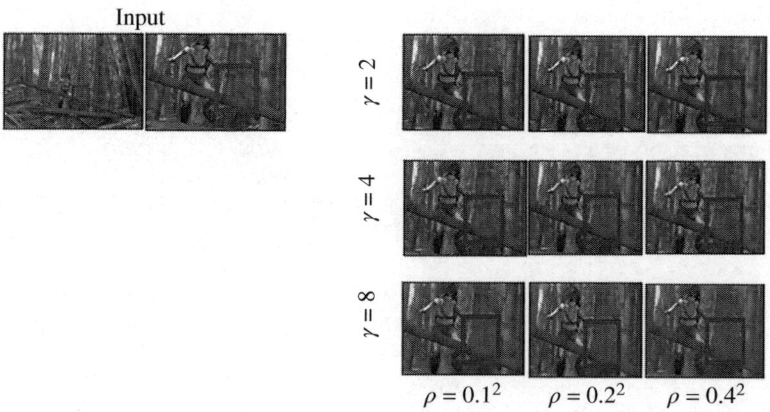

Figure 4.2 Results obtained by the robust WGF with various values of parameters r and ρ. A smaller value of r and ρ improves the robustness of the WGF. *Source:* Sana Rao et al., (2021) / with permission of Elsevier.

Table 4.1 The quantitative evaluation of two edge-preserving filters. The higher values (in bold) represent the higher quality of a filter.

	PSNR					
	GF			WGF		
ρ	$r = 2$	$r = 4$	$r = 8$	$r = 2$	$r = 4$	$r = 8$
0.1^2	19.73	18.85	18.06	**21.22**	**21.05**	**20.82**
0.2^2	19.10	18.05	17.16	**20.94**	**20.53**	**19.99**
0.4^2	19.47	18.51	17.66	**21.16**	**20.91**	**20.59**

performance. The PSNR values are reported in Table 4.1. It can be seen that with the same parameters, the WGF performs better than the GF.

4.3 Robust NLTV-L^1+WGF for Optical Flow Estimation

In this section, we propose a robust NLTV-L^1+WGF method to estimate optical flow. We include the WGF in the NLTV-L^1 for the optical flow estimation. The objective function of the proposed NLTV-L^1+WGF optical flow estimation method can be written as:

$$E(u, v, O, a_i, b_i) = E_{\text{Data}}(u, v) + E_{\text{Smoothness}}(u, v) + E_{NL}(u, v) + E(O, a_i, b_i). \qquad (4.7)$$

The objective function in Eq. (4.7) contains a classical function, a nonlocal term, and a robust WGF. The combination of these items optimizes the flow field by using the MF in the coarse-to-fine process.

4.3.1 Formulation of the Robust WGF for Optical Flow Estimation

For consecutive image frames, let $O_k = (u, v)^T$ indicates the input flow field, where $O:\Omega \rightarrow \mathbb{R}$ denotes the optical flow between two components (u,v), and Ω indicates the image domain. Then, we introduce an auxiliary flow field $\widehat{O}_k = (\widehat{u}, \widehat{v})^T$ that acts as a guidance flow field in the robust WGF. The formulation can be defined as:

$$\widehat{O}_k = a_i O_k + b_i, \tag{4.8}$$

where $\widehat{O}_k = (\widehat{u}_k, \widehat{v}_k)^T$ is the linear transformation of the guidance optical flow $O_k = (u_k, v_k)^T$ at the pixel k.The coefficients $a_i = \left(a_i^u, a_i^v\right)^T$ and $b_i = \left(b_i^u, b_i^v\right)^T$, where $\left(a_i^u, b_i^u\right)$ and $\left(a_i^v, b_i^v\right)$ correspond to the horizontal and vertical components of the flow field O_k, respectively. The values of these linear coefficients can be estimated by minimizing the cost function. Denote the input flow field as $\tilde{O}_k = (\tilde{u}, \tilde{v})^T$. The coefficients a_i and b_i can be estimated as:

$$E(O, a_i, b_i) = \sum_{k \in \omega_i} \left[(a_i O_k + b_i - \tilde{O}_k)^2 + \frac{\rho}{\psi_O(i)} a_i^2 \right]. \tag{4.9}$$

All pixels in the guidance flow field O_k are used in the computation of $\psi_O(i)$. Moreover, the weight function $\psi_O(i)$ measures the edge strength of the pixel i concerning the whole guidance flow field O_k. a_i and b_i can be computed as:

$$a_i = \frac{\frac{1}{N} \sum\limits_{k \in \omega_i} O_k \tilde{O}_k - \overline{O}_i \bar{\tilde{O}}_i}{\sigma_i^2 + \frac{\rho}{\psi_O(i)}} \tag{4.10}$$

$$b_i = \bar{\tilde{O}}_i - a_i \overline{O}_i, \tag{4.11}$$

where σ_i^2 indicates the variance of the guidance flow field O_k in ω_i. $\overline{O}_i$ and $\bar{\tilde{O}}_i$ are the means of the guided flow field O_k and the guided input flow field $\tilde{O}_k$, respectively.

Note that a pixel may belong to different windows. Therefore, the average strategy of the WGF for estimating optical flow can be expressed as:

$$\widehat{O}_k = \frac{1}{N} \sum_{k \in \omega_i} (a_k O_k + b_k), \tag{4.12}$$

which can be rewritten as:

$$\widehat{O}_k = \overline{a}_k O_k + \overline{b}_k, \tag{4.13}$$

where $\overline{a}_k = \frac{1}{N} \sum\limits_{i \in \omega_k} a_i$, and $\overline{b}_k = \frac{1}{N} \sum\limits_{i \in \omega_k} b_i$ are the mean values of a_i and b_i in different filtering windows.

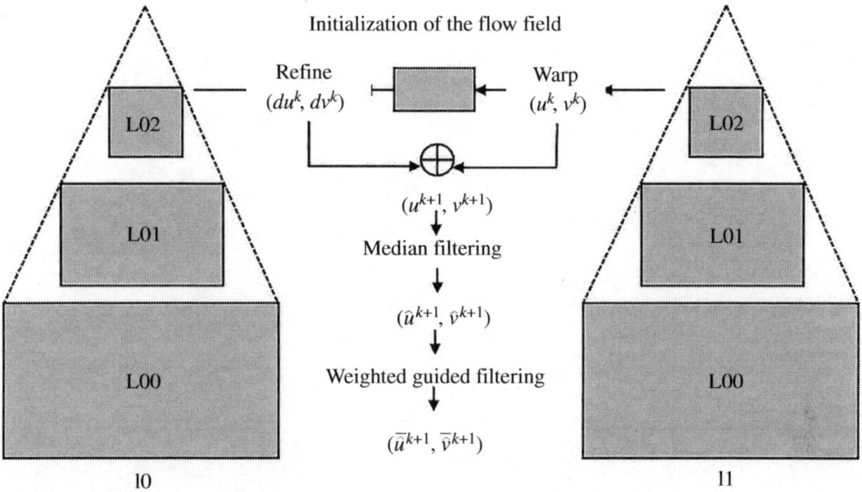

Figure 4.3 The incorporation of the WGF into the coarse-to-fine process to optimize the optical flow.

4.3.2 Optimization of the NLTV-L^1+WGF Optical Flow Estimation

We estimate the optical flow by employing a robust WGF with the MF operation to optimize the NLTV-L^1 optical flow, using the coarse-to-fine process. In some state-of-the-art optical flow estimation methods, filtering during the optical flow optimization can effectively remove outliers and make the optical flow estimation more accurate and robust. Figure 4.3 shows the optimization process, where the pyramid levels are labeled as L00, L01, etc. During the coarse-to-fine process, we assume that the total number of levels in the pyramid is n. Starting from the coarsest level, the flow is refined at each level iteratively.

4.4 Experiments and Analysis

We conduct a series of experiments to verify the robustness of the proposed NLTV-L^1+WGF method. Moreover, we compare the proposed method with several other state-of-the-art optical flow estimation methods.

4.4.1 The Influence of Parameters

In this section, we study the influence of parameters on the performance of our proposed NLTV-L^1+WGF method. We evaluate the influence of parameters ρ and r on the accuracy of flow field results, as shown in Figures 4.4 and 4.5. The value of ρ affects the accuracy of the flow field (e.g., the smaller the value of ρ is, the

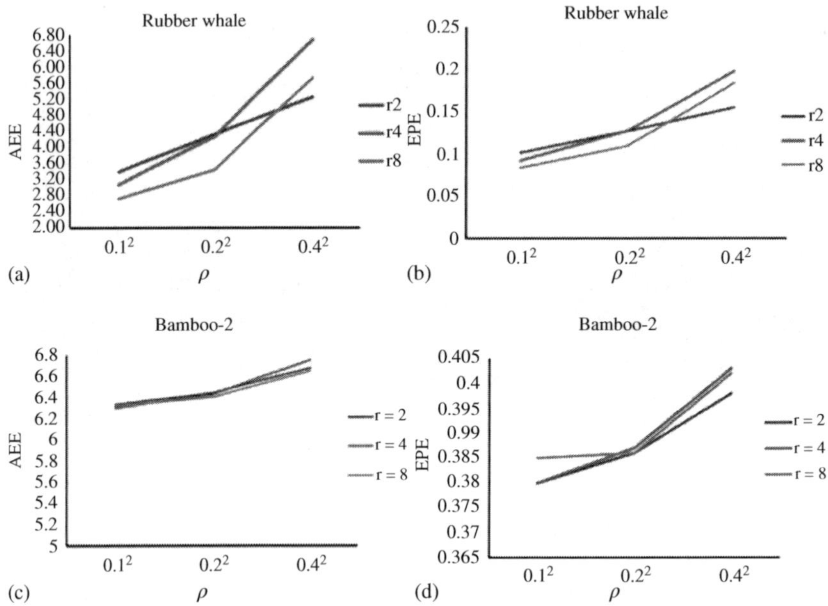

Figure 4.4 The AAE and EPE obtained by the proposed NLTV-L^1+WGF method on the RubberWhale sequence (of the Middlebury dataset) and the Bamboo-2 sequence (of the MPI-Sintel dataset) using various values of ρ and r in the proposed NLTV-L^1+WGF method.

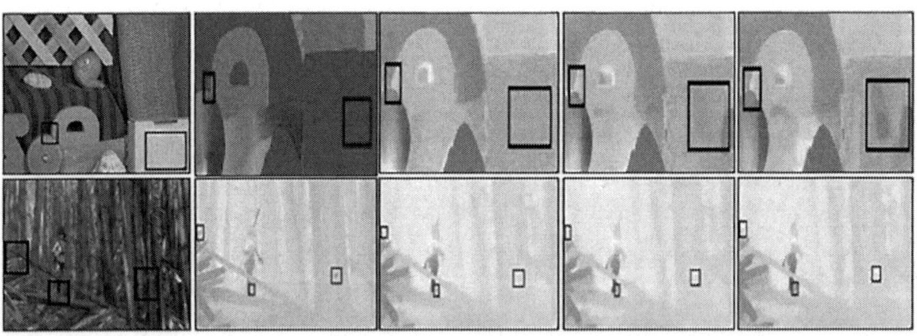

Figure 4.5 The input image frames (in the left), the ground-truths of the optical flow (the second column), and the results obtained by the proposed NLTV-L^1+WGF method on the RubberWhale sequence of the Middlebury dataset (the first row) and the Bamboo-2 sequence of the MPI-Sintel dataset (the second row) with $\rho = 0.1^2, 0.2^2, 0.4^2$, respectively (the third column to the fifth column). *Source:* Sana Rao et al., (2021) / with permission of Elsevier.

more accurate the estimated flow field is). To examine the influence of various values for ρ and r on the performance of the proposed method, we summarize the Angular Error (AAE) and End-point-error (EPE) in Figure 4.4. The marked areas in Figure 4.5 show the smoothness around the object boundaries.

4.5 Results

4.5.1 Results on the Middlebury Dataset

We evaluate the performance of the proposed NLTV-L^1+WGF optical flow estimation method on the Middlebury dataset. We compare the proposed method with several other state-of-the-art optical flow estimation methods (including the RNLTV-L^1 [14], PMC [17], JOF [3], AGFOF [18], NLP [19], RNLOD [1], NNF+EAC [20], and TVG [21] methods). The JOF method employs an additional mutual-structure GF to solve the inconsistency problem in optical flow estimation, and it refines the TV-L^1 optical flow estimation method using a joint filter. The AGFOF method employs the GF to the optical flow estimation method, which is robust to preserve motion boundaries in normal scenarios. The NLP method employs a nonlocal propagation filter to the optical flow estimation method to deal with the problem of ill-defined edges near the motion boundaries. The RNLOD method exploits a robust nonlocal term to improve the optical flow estimation.

Table 4.2 shows the AAE and EPE results obtained by the nine competing methods on several image sequences of the Middlebury dataset. The AAE and EPE values obtained by the proposed NLTV-L^1+WGF method is reduced for four out of six image sequences because of the edge-preserving WGF.

Figure 4.6 shows the estimated flow field results obtained by the proposed method and the other state-of-the-art methods on the Venus and Grove sequences of the Middlebury dataset, where the black boxes indicate several regions containing motion boundaries. The results show that the proposed NLTV-L^1+WGF method can effectively preserve edges. The detailed comparisons show that the proposed NLTV-L^1+WGF method accurately preserves the edges of the tree branch in the Grove sequence. Moreover, the results obtained by the proposed optical flow estimation method on the Venus sequence are visually close to the ground-truth.

4.5.2 Results on the MPI-Sintel Dataset

In this section, we evaluate the proposed NLTV-L^1+WGF method on the MPI-Sintel dataset by conducting a comparison with several state-of-the-art methods. Table 4.3 shows the obtained AAE and EPE errors by the competing

Table 4.2 The results obtained by the nine competing methods on six complex image sequences of the Middlebury dataset. The bold values represent the lowest AAE and EPE errors.

	Dimetrodon		Yosemite		Grove2		Grove3		Venus		RubberWhale	
Method	**AAE**	**EPE**	**AAE**	**EPE**	**AAE**	**EPE**	**AAE**	**EPE**	**AAE**	**EPE**	**AAE**	**EPE**
PMC (SPIC 2022) [17]	2.56	**0.13**	2.45	0.14	2.83	0.18	4.98	**0.41**	4.14	0.26	2.69	0.09
JOF (TMM 2020) [3]	2.37	0.15	3.12	0.19	2.70	0.15	6.18	0.68	5.10	0.37	2.33	0.07
AGFOF (SP 2016) [18]	2.56	0.13	2.48	0.13	2.80	0.16	**4.45**	0.44	5.41	0.32	3.01	0.09
NLP (SPIC 2021) [19]	2.48	0.21	2.46	0.12	2.66	0.15	4.85	0.47	4.36	0.25	2.35	**0.06**
NNF+EAC (PR 2017) [20]	2.64	0.13	2.36	0.14	2.70	0.19	5.06	0.48	4.96	0.29	4.26	0.12
RNLOD (TIP 2017) [1]	3.45	0.35	3.00	0.16	2.68	0.15	6.07	0.62	4.38	0.27	2.39	0.07
TVG (ICBD 2018) [21]	2.92	0.30	2.32	0.13	2.79	0.19	7.09	0.69	5.28	0.30	4.60	0.15
RNLTV-L^1 (MTAP 2021) [14]	2.80	0.37	2.25	0.13	2.68	0.15	4.98	**0.41**	4.09	0.22	**2.28**	**0.06**
NLTV-L^1 + WGF [22]	**2.36**	0.15	**2.23**	**0.13**	**2.55**	**0.14**	4.92	**0.41**	**3.82**	**0.23**	**2.28**	**0.06**

Table 4.3 The results obtained by the seven competing methods on five complex image sequences of the MPI-Sintel dataset. The bold values represent the lowest AAE and EPE errors.

	Alley-1		Bamboo		Mountain		Market		Cave	
Method	**AAE**	**EPE**	**AAE**	**EPE**	**AAE**	**EPE**	**AAE**	**EPE**	**AAE**	**EPE**
RNLTV-L^1 (MTAP 2021) [14]	2.39	0.25	4.17	0.22	8.80	1.48	12.90	2.47	5.18	1.73
TVG (ICBD 2018) [21]	2.44	**0.19**	4.25	0.21	20.55	2.02	14.31	2.89	5.28	**0.30**
DIOF (ECCV 2018) [2]	2.40	0.22	7.25	1.55	12.05	1.83	18.43	2.85	10.53	2.05
NNF+EAC (PR 2017) [20]	**2.18**	0.20	4.27	0.23	12.50	1.89	13.02	2.31	**5.03**	1.53
RNLOD (TIP 2017) [1]	2.99	0.37	4.39	0.27	17.77	2.34	25.97	6.38	17.26	3.43
NLP (SPIC 2021) [19]	2.97	0.35	4.35	0.24	17.70	2.30	25.94	6.30	7.61	0.50
NLTV-L^1+WGF [22]	2.30	0.20	**4.01**	**0.21**	**7.56**	**1.36**	**12.87**	**2.30**	5.04	1.65

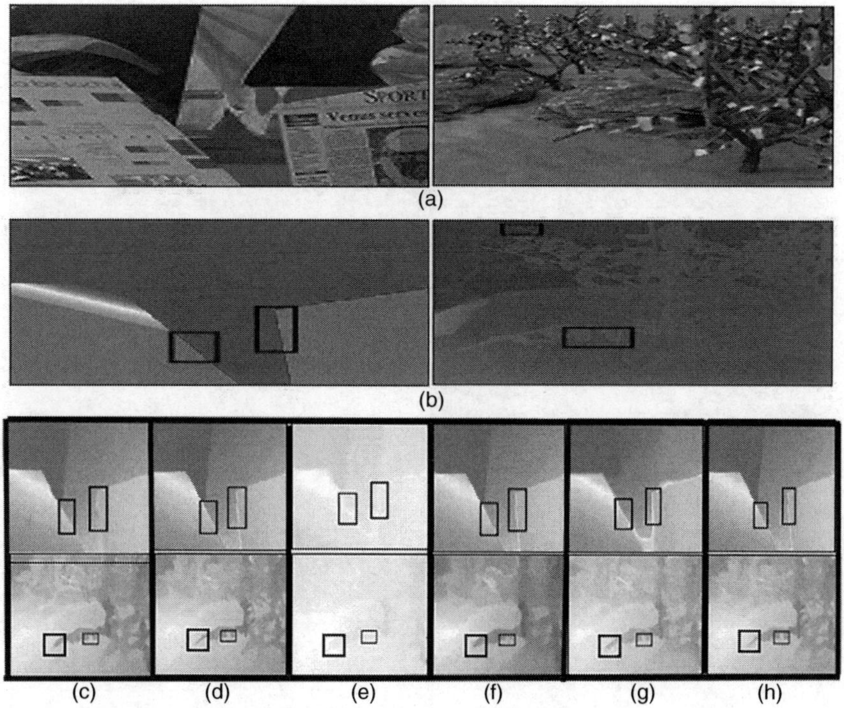

Figure 4.6 Visual comparison of the optical flow estimation results obtained by the proposed NLTV-L^1+WGF method and several other competing methods on two complex image sequences of the Middlebury dataset, where the black boxes indicate the problem of blurring motion boundaries. (a) The input image frames. (b) The ground-truths of the optical flow. (c–h) The results obtained by the AGFOF, NLP, TVG, JOF, RNLTV-L^1, and NLTV-L^1+WGF methods, respectively. *Source:* Sana Rao et al., (2021) / with permission of Elsevier.

methods on several image sequences of the MPI-Sintel dataset. The proposed NLTV-L^1+WGF method outperforms the other state-of-the-art optical flow estimation methods mainly because the proposed method performs better near motion boundaries. However, the weighting parameters may not be the best choice for the Market sequence.

For the visual comparison of the results obtained by the proposed NLTV-L^1+WGF method and several other competing methods, we show the estimated optical flow on two complex image sequences of the MPI-Sintel dataset in Figure 4.7, where the black boxes indicate motion boundaries. The result shows that the proposed method preserves the edges near the motion boundaries. The qualitative comparison indicates that the proposed method has the benefit of

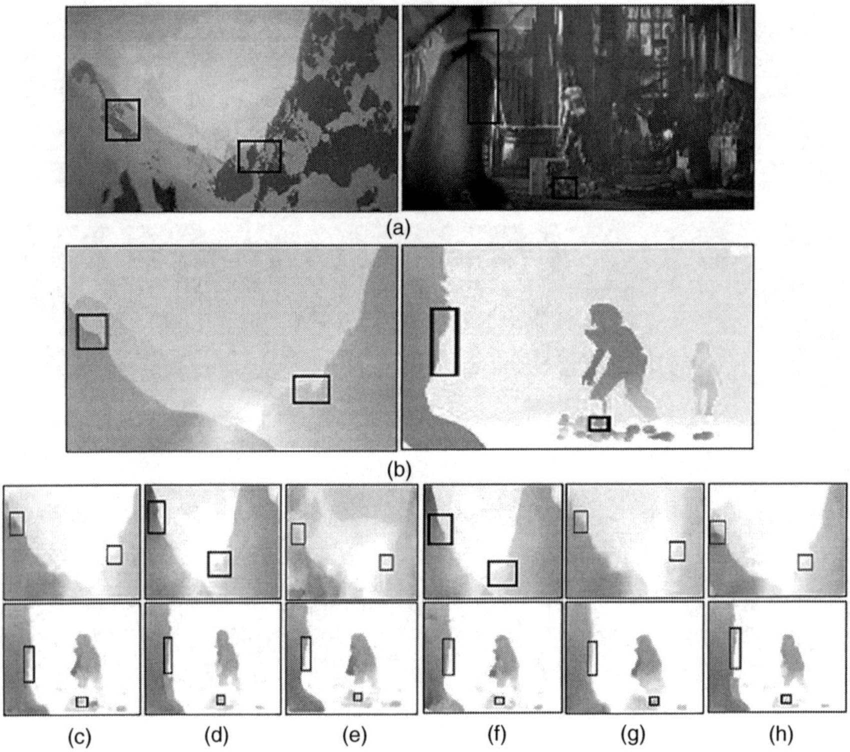

Figure 4.7 Visual comparison of the optical flow estimation results obtained by the proposed NLTV-L^1+WGF method and several other competing methods on two complex image sequences of the Middlebury dataset, where the black boxes indicate the problem of blurring motion boundaries. (a) The input image frames. (b) The ground-truths of the optical flow. (c–h) The results obtained by the AGFOF, NLP, TVG, JOF, RNLTV-L^1, and NLTV-L^1+WGF methods, respectively. *Source:* Sana Rao et al., (2021) / with permission of Elsevier.

edge preservation. Figure 4.7 shows that the edges of the tiny objects (e.g., the apple on the floor) are effectively preserved by the proposed method.

4.5.3 Results on the Foggy Zurich Dataset

The proposed NLTV-L^1+WGF method is evaluated on a more challenging outdoor dataset. Figure 4.8 shows that the proposed NLTV-L^1+WGF method achieves satisfactory performance compared to the other state-of-the-art optical flow estimation methods. For a detailed comparison, we show the magnified views within the black boxes in Figure 4.8. It can be noticed that the sidewalk (2) is missed in the results obtained by the TVG and DIOF methods, partially recovered in the results

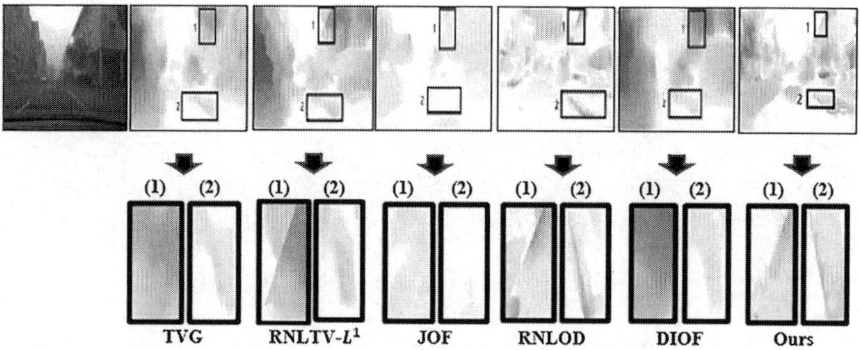

Figure 4.8 The visual comparison of results obtained by the proposed optical flow estimation method and several other competing methods on the Foggy Zurich image sequence with the magnified view of the black boxes (i.e., (1) the house shade and (2) the sidewalk). *Source:* Sana Rao et al., (2021) / with permission of Elsevier.

obtained by the JOF method, but almost entirely restored in the results obtained by the proposed method. The RNLOD method and the proposed NLTV-L^1+WGF method can also recover the edges of the house shade (1).

4.5.4 Ant Motion Estimation

We evaluate the proposed method on a more complex ant's motion dataset [23]. Figure 4.9 shows the results obtained by the proposed NLTV-L^1+WGF method on the ant motion image sequences. This dataset is complex because most of the ants move quickly. To further analyze the motion of ants, we focus on a specific ant, as shown in Figure 4.10. Examples of the robust optical flow estimation for the ants' interaction, occlusion, and no motion are shown in Figure 4.11.

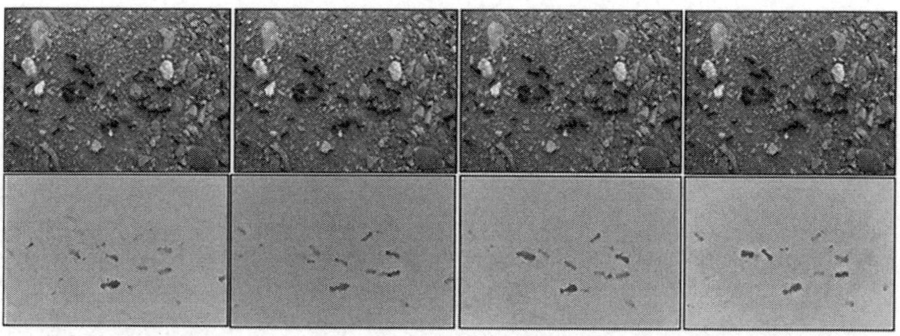

Figure 4.9 The motion estimation of ants obtained by the proposed NLTV-L^1+WGF method under the noisy outdoor environment. *Source:* Sana Rao et al., (2021) / with permission of Elsevier.

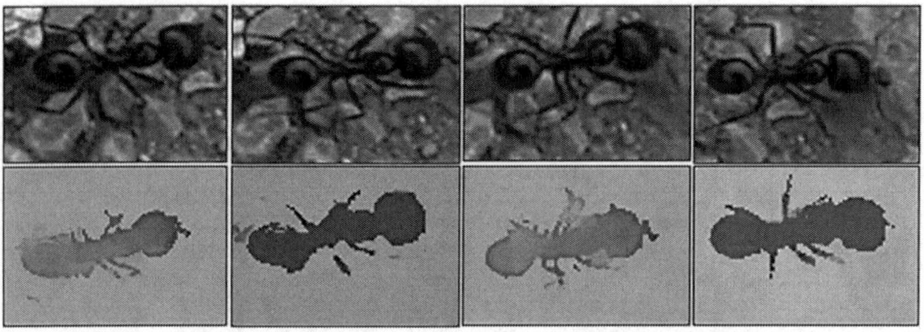

Figure 4.10 Amplified view of an ant's motion estimation obtained by the proposed NLTV-L^1+WGF method in different lighting conditions. *Source:* Sana Rao et al., (2021) / with permission of Elsevier.

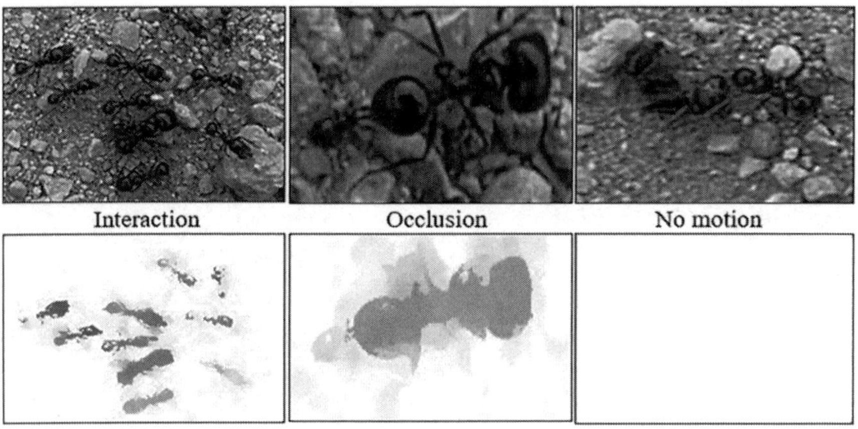

Figure 4.11 The optical flow of ants' motion estimated by the proposed NLTV-L^1+WGF method in different scenarios (i.e., interaction, occlusion, and no motion). *Source:* Sana Rao et al., (2021) / with permission of Elsevier.

4.6 Conclusion

In this chapter, a NLTV-L^1+WGF method is proposed to preserve edges and motion boundaries of moving objects. We introduce the WGF to optical flow estimation. Furthermore, we combine WGF with NLTV-L^1 to optimize the optical flow using the coarse-to-fine process. The proposed method is evaluated on several challenging datasets and compared with several other state-of-the-art optical flow estimation methods. The results show that the proposed method can estimate an accurate and robust optical flow because of its edge-preserving ability.

Despite the success in terms of the robustness of the proposed NLTV-L^1+WGF optical flow estimation method, the WGF optimization is still not robust enough because it neglects the directions of edges in the image. In the next chapter, to improve the robustness of optical flow estimation, we propose to combine the WGF with a nonlocal steering kernel for the optical flow estimation. The WGF can preserve the edges of an image, and the nonlocal steering kernel can leverage the edge direction that is robust for the optical flow estimation.

Examples with Solutions

Edge-aware Weight Computation in WGF

Problem Statement:
Given a 5×5 grayscale image patch, compute the edge-aware weight function $\psi_I(i)$ for the central pixel. Assume dynamic range L = 255, $\lambda = (0.001\text{L})^2$, and provide interpretation of results.

Solution Steps:

1) Write down the given 5×5 matrix of intensity values.

$$\begin{bmatrix} 100 & 102 & 104 & 106 & 108 \\ 100 & 103 & 105 & 107 & 110 \\ 98 & 101 & 150 & 109 & 111 \\ 97 & 100 & 103 & 106 & 108 \\ 96 & 99 & 102 & 105 & 107 \end{bmatrix}$$

- The central pixel has intensity 150 (row 3, column 3).
- We will calculate the variance around this pixel using a 3×3 window.

2) Extract the 3×3 window around central pixel

$$\begin{bmatrix} 103 & 105 & 107 \\ 101 & 150 & 109 \\ 100 & 103 & 106 \end{bmatrix}$$

3) Compute local variance for each 3×3 window.
   - Mean intensity of this window:

$$\mu = \frac{1}{9}(103 + 105 + 107 + 101 + 150 + 109 + 100 + 103 + 106)$$

$$\mu = \frac{984}{9} \approx 109.33$$

- Variance formula:

$$\sigma^2 = \frac{1}{9}\sum (I_k - \mu)^2$$

Now compute each squared difference:

- (103–109.33)2 ≈ 40.1
- (105–109.33)2 ≈ 18.8
- (107–109.33)2 ≈ 5.4
- (101–109.33)2 ≈ 69.4
- (150–109.33)2 ≈ 1655.1
- (109–109.33)2 ≈ 0.1
- (100–109.33)2 ≈ 87.1
- (103–109.33)2 ≈ 40.1
- (106–109.33)2 ≈ 11.1

Sum of squared differences = 1927.2

$$\sigma^2 = \frac{1927.2}{9} \approx 214.13$$

4) Normalize variance.
We now compute the **average variance** across the patch.
   - Suppose, after similar calculations for other windows, the **mean variance** = 50 (this is a typical value for nonedge regions in this patch).
     So we will use:

$$\frac{1}{N}\sum_{i'=1}^{N} \sigma_{I,1}^2(i') = 50$$

   - Calculate $\psi_I(i)$ for the central pixel.
     Formula:

$$\psi_I(i) = \frac{\frac{1}{N}\sum_{i'=1}^{N} \sigma_{I,1}^2(i') + \lambda}{\sigma_I^2(i) + \lambda}$$

$$\psi_I(i) = \frac{50 + 0.065}{214.13 + 0.065}$$

$$\psi_I(i) \approx \frac{50.065}{214.195} \approx 0.234$$

   - Interpret:
     - If $\psi_I(i)$ is **close to 1** → flat/smooth region (high weight).
     - If $\psi_I(i)$ is **much less than 1** → edge region (low weight, edge preserved).

$$\psi_I(i) \approx 0.23 \ll 1$$

Answer:
Step-by-step calculations show that variance around the central pixel is high, leading to $\psi_I(i) > 1$. Hence, WGF will assign larger weight, preserving the edge.

Minimization of WGF Cost Function

Problem Statement:
Minimize the cost function $E(a_i, b_i)$ for a given 3×3 window with guidance intensities I_k and input values X_k.

Solution Steps:

1) Define the 3×3 guidance window I and the corresponding input X.

$$I = \begin{bmatrix} 60 & 62 & 65 \\ 61 & 63 & 66 \\ 64 & 67 & 69 \end{bmatrix}, X = \begin{bmatrix} 58 & 60 & 63 \\ 59 & 61 & 64 \\ 62 & 65 & 67 \end{bmatrix}$$

Thus, we have 9 pairs of (I_k, X_k).

2) Write the cost function $E(a_i, b_i)$.
The cost function is:

$$E(a_i, b_i) = \sum_{k=1}^{9} \left((a_i, I_k + b_i - X_k)^2 + \varepsilon a_i^2\right)$$

This means for every pixel in the local window, we measure how well a linear function of the guidance image $(a_i I_k + b_i)$ approximates the input image X_k.

3) Differentiate w.r.t. a_i and b_i.

$$\frac{\partial E}{\partial a_i} = \sum_{k=1}^{9}(2(a_i I_k + b_i - X_k)I_k) + 2\varepsilon a_i = 0$$

4) Solve the resulting linear equations.
We simplify into two equations:
   1) Equation (1):

$$a_i \sum I_k^2 + b_i \sum I_k - \sum X_k I_k + \varepsilon a_i \cdot 9 = 0$$

$$a_i \left(\sum I_K^2 + 9\varepsilon\right) + b_i \sum I_k = \sum X_k I_k$$

   2) Equation (2):

$$a_i \sum I_k + 9b_i = \sum X_k$$

$$a_i \approx 1.40, \quad b_i \approx -27.64$$

Answer:
The optimal coefficients a_i and b_i are derived, showing how WGF locally adapts slope and intercept to preserve edges while smoothing noise.

Parameter Influence on WGF

Problem Statement:
Investigate how different values of ρ affect the PSNR of the filtered image. Compare cases for $\rho = 0.1^2, 0.2^2$, and 0.4^2.

Solution Steps:

1) Tabulate PSNR values for GF and WGF.
   Assume we test on a standard synthetic dataset (e.g., a grayscale image with edges + flat regions) using different **filter radii *r*** ($r = 4, 8, 16$).
2) Compare improvements at different radii r.

   At small ρ (0.1^2):
   - WGF shows ~**2.5 dB gain** over GF.
   - Best performance since edge sharpness is preserved and noise is reduced.

   At medium ρ (0.2^2):
   - PSNR drops slightly, but WGF still outperforms GF by **2–2.5 dB**.
   - Balance between smoothness and sharpness.

   At large ρ (0.4^2):
   - Both GF and WGF suffer from blurring (lower PSNR).
   - WGF still provides a small edge-preservation advantage (~1.5–2 dB gain).
3) Interpret trends.

Answer:

- For small ρ, WGF outperforms GF significantly.
- Larger ρ increases blurring (lower PSNR).
- WGF consistently preserves edges better than GF.

Applying WGF in Optical Flow Estimation

Problem Statement:
Given two synthetic consecutive frames (10×10 pixels, moving square object), compute optical flow using NLTV-L^1+WGF.

Solution Steps:

1) Define brightness constancy assumption.
2) Formulate the NLTV-L^1 energy functional with WGF.
3) Refine the flow.
4) Visualize the resulting flow vectors.

Answer:
Results show that NLTV-L^1+WGF preserves motion boundaries at the edges of the square object, avoiding halo artifacts common in GF.

Coarse-to-fine Optimization

Problem Statement:
Demonstrate the coarse-to-fine flow estimation process with pyramid levels L00, L01, L02 for a 32×32 sequence.

Solution Steps:

1) Construct Gaussian pyramid.
2) Initialize flow at coarsest level.
3) Iteratively refine using WGF.
4) Show intermediate results at each level.

Answer:
The pyramid reduces computational cost and helps capture large motions at coarse levels, refined later with WGF to preserve edges.

References

1 Zhang, C., Chen, Z., Wang, M. et al. (2017). Robust non-local tv-l1 optical flow estimation with occlusion detection. *IEEE Transactions on Image Processing* 26 (8): 4055–4067.

2 Li, R., Tan, R.T., and Cheong, L.F. (2018). Robust optical flow in rainy scenes. *European Conference on Computer Vision,* 8–14 September 2018, Munich, Germany. 288–304.

3 Zhang, C., Ge, L., Chen, Z. et al. (2019). Refined tv-l1 optical flow estimation using joint filtering. *IEEE Transactions on Multimedia* 22 (2): 349–364.

4 Sun, D., Roth, S., and Black, M.J. (2014). A quantitative analysis of current practices in optical flow estimation and the principles behind them. *International Journal of Computer Vision* 106 (1): 115–137.

5 Mei, L., Lai, J., Xie, X. et al. (2019). Illumination-invariance optical flow estimation using weighted regularization transforms. *IEEE Transactions on Circuits and Systems for Video Technology* 30 (2): 495–508.

6 Horn, B.K. and Schunck, B.G. (1981). Determining optical flow. *Artificial Intelligence* 17 (1–3): 185–203.

7 Li, Z., Zheng, J., Zhu, Z. et al. (2014). Weighted guided image filtering. *IEEE Transactions on Image Processing* 24 (1): 120–129.

8 Wu, J., Wang, X., Chen, Z., Zhang, C. (2018). Robust variational optical flow algorithm based on rolling guided filtering. *International Congress on Image and Signal Processing, BioMedical Engineering and Informatics*, Oct 13–15, 2018, China. 1–6.

9 He, K., Sun, J., and Tang, X. (2012). Guided image filtering. *IEEE Transactions on Pattern Analysis and Machine Intelligence* 35 (6): 1397–1409.

10 Navarro, J. and Buades, A. (2017). Robust and dense depth estimation for light field images. *IEEE Transactions on Image Processing* 26 (4): 1873–1886.

11 George, G., Oommen, R.M., Shelly, S. et al. (2018). A survey on various median filtering techniques for removal of impulse noise from digital image. *Conference on Emerging Devices and Smart Systems*, 2–3 March, 2018, Tiruchengode, India. 235–238.

12 Liu, W., Chen, X., Shen, C. et al. (2017). Robust guided image filtering, arXiv preprint:1703.09379.

13 Qiu, X., Li, M., Zhang, L., and Yuan, X. (2019). Guided filter-based multi-focus image fusion through focus region detection. *Signal Processing: Image Communication* 72 (1): 35–46.

14 Rao, S. (2021). A framework for robust motion estimation and segmentation in adverse outdoor conditions. *Multimedia Tools and Applications*, 1–21.

15 Tomasi, C. and Manduchi, R. (1998). Bilateral filtering for gray and color images. *IEEE International Conference on Computer Vision,* 4–7 Jan, 1998, Bombay, India. 839–846.

16 Schmidt, B.E., Skiba, A.W., Hammack, S.D. et al. (2021). High-resolution velocity measurements in turbulent premixed flames using wavelet-based optical flow velocimetry. *Proceedings of the Combustion Institute* 38 (1): 1607–1615.

17 Zhang, C., Feng, C., Chen, Z. et al. (2022). Parallel multiscale context-based edge-preserving optical flow estimation with occlusion detection. *Signal Processing: Image Communication* 101 (1): 116545–116560.

18 Tu, Z., Poppe, R., and Veltkamp, R.C. (2016). Adaptive guided image filter for warping in variational optical flow computation. *Signal Processing* 127 (1): 253–265.

19 Dong, C., Wang, Z., Han, J. et al. (2021). A non-local propagation filtering scheme for edge- preserving in variational optical flow computation. *Signal Processing: Image Communication* 93 (1): 116125–116143.

20 Tu, Z., Xie, W., Cao, J. et al. (2017). Variational method for joint optical flow estimation and edge-aware image restoration. *Pattern Recognition* 65 (1): 11–25.

21 Zhang, Y., Fan, H., Zheng, J., and Zhang, C. (2018). Improving the optical flow accuracy based on the total variation of local-global method. *IEEE International Conference on Big Data, Big Data,* Dec 10–13, 2018, WA, USA. 4658–4664.

22 Rao, S. and Wang, H. (2021). Robust optical flow estimation via edge preserving filtering. *Signal Processing: Image Communication* 96 (1): 116295–116309.

23 Cao, X., Guo, S., Lin, J. et al. (2020). Online tracking of ants based on deep association metrics: Method, dataset and evaluation. *Pattern Recognition* 103 (1): 107220–107233.

5

Optical Flow Estimation via Weighted Guided Filter with Nonlocal Steering Kernel

5.1 Introduction

During the past few decades, several optical flow estimation methods have been proposed to handle the abovementioned problems [1, 2]. The variational [3–5], total variational [6, 7], nonlocal total variational (NLTV) [8–11], and convolutional neural network-based (CNN-based) methods [12–17] are considered the important techniques for optical flow estimation. Most optical flow estimation methods belong to the variational methods. Those variational-based optical flow estimation methods have been modified and extended in two different ways: (1) an improvement in the data term to make the optical flow more robust under illumination changes, and (2) an improvement in the regularization term to handle the complex motion boundaries.

To address the problem of edge blurring near motion boundaries and ensure the robustness of optical flow, we propose to combine the weighted guided filter with a nonlocal steering kernel (WGF-NLSK) for the optical flow estimation method. A NLSK contains both the neighboring similarity and the local spatial support to address the problem of motion boundaries in the optical flow estimation. We propose to employ a NLSK in the WGF as a weighted average strategy to improve the filter's behavior.

5.1.1 Motivation for the Proposed Approach

This limitation in handling sharp motion boundaries motivates the development of the WGF-NLSK, which this chapter will discuss in detail. The NLSK builds on the strength of nonlocal means, a method introduced for image denoising. Nonlocal means rely on comparing the similarity between patches in an image to preserve fine structures. By combining this nonlocal approach with the edge-preserving properties of the WGF, we can overcome the challenge of preserving motion boundaries while maintaining computational efficiency.

Robust Optical Flow Estimation in Outdoor Environments, First Edition. Sana Rao and Rao Kashif.

The proposed method aims to achieve several key objectives:

1) **Improve Edge Preservation**: Through the introduction of the NLSK, we aim to enhance the handling of edge regions where the motion changes drastically, such as around moving objects that occlude one another.
2) **Reduce Halo Artifacts**: One common issue with guided filtering (GF) in optical flow is the appearance of halos or blurred edges, especially around areas with large illumination variations. The NLSK enhances the regular GF's ability to handle these variations more effectively.
3) **Adaptability to Real-World Conditions**: The combination of the WGF and NLSK ensures that the method can adapt to various adverse outdoor conditions, such as varying lighting, occlusion, fog, and rain, making it particularly suited for real-world applications like autonomous driving and video surveillance.

The following sections will explore these topics in greater detail. We begin by revisiting the fundamental principles of the WGF and then introduce the NLSK as an enhancement to traditional methods. After detailing the optimization process, we will present a comprehensive evaluation of the proposed method on several benchmark datasets, analyzing its performance in a variety of scenarios.

5.2 The WGF with a Nonlocal Steering Kernel for Optical Flow Estimation

5.2.1 Overview of the Weighted Guided Filter (WGF)

The WGF is an extension of the traditional GF, which itself is a form of edge-preserving filter. Unlike conventional smoothing filters, which tend to blur edges, the GF can maintain sharp edges while denoising an image or video. It operates by using a guidance image (which can be the input image itself or another related image) to control the filtering process.

In the case of optical flow estimation, the WGF helps smooth the flow field while preserving motion boundaries. The core idea behind the WGF is to compute linear coefficients that describe the relationship between a guidance image I and the input optical flow field. These coefficients are then used to compute a refined flow field that smooths out noise while retaining important structural details in the motion.

The filtering process can be expressed mathematically as:

$$q_i = a_k I_i + b_k, \tag{5.1}$$

where I_i is the guidance image, q_i is the filtered output, and a_k and b_k are linear coefficients calculated based on the local structure of the image. These coefficients are designed to preserve edges by adapting to the intensity gradients in the guidance image.

However, while the WGF performs well in many cases, it struggles in scenarios with occlusions and large illumination changes. This limitation occurs because the WGF's edge preservation relies on local intensity gradients, which may not always correspond to the true motion boundaries in an optical flow field.

5.2.2 Introduction to the Nonlocal Steering Kernel (NLSK)

The NLSK addresses these limitations by incorporating a more global understanding of the image's structure. Nonlocal means methods, which form the basis of the NLSK, consider not just the local pixel neighborhood but also distant pixels with similar intensities or textures. By comparing patches rather than individual pixels, nonlocal methods can capture more complex structures and textures in an image.

The NLSK builds on this idea by using both local and nonlocal information to steer the filtering process. The kernel assigns greater weights to pixels that are both nearby and structurally similar to the pixel being processed. This approach ensures that the filter takes into account global image information, which helps in preserving motion boundaries even in the presence of occlusions or large illumination changes.

Mathematically, the NLSK is formulated as follows:

$$\psi_I(i) = \frac{1}{N}\sum_{k=1}^{N}\frac{\sigma_{I,1}^2(i)+\lambda}{\sigma_{I,1}^2(k)+\lambda}. \tag{5.2}$$

Due to the edge-aware ability of the weight function in Eq. (5.2), the WGF outperforms the other edge-preserving filters in many applications. The WGF adopts an averaging strategy (via the mean filter) to deal with overlapping windows. The averaging strategy neglects the directions of the edges. As a result, the edges are prone to be blurred in this process.

We propose to employ a WGF-NLSK to overcome this problem. We consider a steering kernel, which is one of the most attractive methods in the image processing field [18, 19]. It contains both the neighborhood similarity and the local spatial support. Additionally, it allocates large weights to locally and structurally similar pixels and central pixels of nonlocal similar patches. A NLSK can be defined as:

$$W_{NLSK}^{u,v}(i) = \frac{W_{NL}\left(x_i, \vec{x}_i^{u,v}\right).k\left(x_i, \vec{x}_i^{u,v}\right)}{\sum_{u=-r}^{P}\sum_{v=0}^{p} W_{NL}\left(x_i, \vec{x}_i^{u,v}\right).k\left(x_i, \vec{x}_i^{u,v}\right)}, \tag{5.3}$$

where

$$W_{NL}(i,j) = \exp\left[-\frac{\|N_i(u) - N_j(u)\|}{\sigma^2}\right], \tag{5.4}$$

and

$$k(x_i, x_j) = \frac{\sqrt{det(C_j)}}{h^2} . \exp\left[\frac{(x_i - x_j)^T C_j (x_i - x_j)}{-2h^2}\right]. \tag{5.5}$$

In Eq. (5.3), i indicates the pixel in the image, and j represents the pixel located in the r sized search window around pixel i. In the nonlocal weight function in Eq. (5.4), $N_i(u)$ and $N_j(u)$ denote the r-sized similarity patch surrounding the pixel i and j, respectively, while the symbol σ denotes the filtering parameter. Equation (5.5) represents a local steering kernel, where $x_j \in N(x_i)$ and C_j is the symmetric covariance matrix that is evaluated by the spatial gradient vectors surrounding x_j. The NLSK in Eq. (5.3) is a combination of the nonsimilarity and the local structure properties, where $\vec{x}_i^{u,v}$ is a pixel located in the patch that shifts pixels u and v for x_i in the horizontal and vertical directions, respectively. The WGF-NLSK can be formulated as:

$$q_i = \sum_{k \in w_i} W_{NLSK}(a_k I_i + b_k), \tag{5.6}$$

where a_k and b_k are the linear coefficients, whose values are determined by minimizing Eq. (5.6). W_{NLSK} is a NLSK as a weight function. Equation (5.6) is reformulated as:

$$q_i = \bar{a}_i I_i + \bar{b}_i, \tag{5.7}$$

where $\bar{a}_i = \sum_{k \in w_i} W_{NLSK} a_k$ and $\bar{b}_i = \sum_{k \in w_i} W_{NLSK} b_k$ are the weighted averages of a_k and b_k, respectively. The flow chart of the proposed NLSK-based WGF is presented in Figure 5.1. Given the input image P and the guidance image I, the WGF-NLSK first computes the edge-aware weight from Eq. (5.2). After obtaining a NLSK W_{NLSK} from Eq. (5.3), it computes the weighted average of a_k and b_k. Finally, it outputs q_i by Eq. (5.7).

To further evaluate the performance of the WGF-NLSK, we compare it with the GF and the WGF on the MPI-Sintel image sequence. In the conducted experiments, the global smoothness parameter h of the NLSK W_{NLSK} is set to 2.4. There

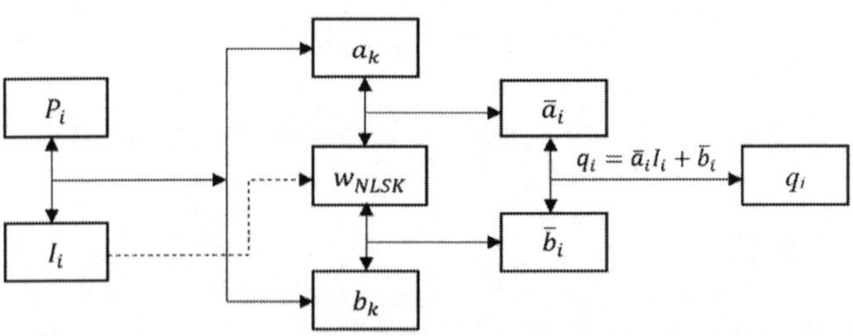

Figure 5.1 A flow chart of the WGF with a non-local steering kernel.

Table 5.1 The quantitative evaluation of three edge-preserving filters. The higher values (in bold) represent the higher quality of a filter.

	PSNR								
	GF			WGF			WGF-NLSK		
ρ	$r = 2$	$r = 4$	$r = 8$	r=2	$r = 4$	$r = 8$	$r = 2$	$r = 4$	$r = 8$
0.1^2	17.45	17.32	17.16	17.56	17.55	17.51	**17.74**	**17.70**	**17.68**
0.2^2	17.30	17.02	16.67	17.55	17.51	17.42	**17.60**	**17.56**	**17.51**
0.4^2	17.16	16.74	16.24	17.52	17.43	17.26	**17.57**	**17.50**	**17.40**

are two other parameters (i.e., the radius r and the smoothing parameter ρ), which are important to be determined in WGF-NLSK. The PSNR values of the results are reported in Table 5.1. It can be noticed that with the same parameters, the WGF-NLSK outperforms GF and WGF.

5.3 Optimization of the Optical Flow Estimation

5.3.1 Coarse-to-Fine Optimization Framework

Optical flow estimation is inherently an optimization problem. The goal is to find the flow field that best describes the motion between two consecutive frames while maintaining smoothness in areas of uniform motion and preserving sharp boundaries where motion discontinuities occur. To solve this optimization problem, we employ a coarse-to-fine optimization framework, which is a common strategy in variational methods for optical flow estimation.

The filtering-based optical flow optimization using the coarse-to-fine process effectively preserves the edges near the motion boundaries and makes the optical flow estimation more robust. However, the WGF may yield the halo artifacts around the motion boundaries under complex outdoor conditions. To overcome the limitations, we estimate the optical flow by combining the robust WGF-NLSK to optimize the NLTV-L^1 optical flow, using the coarse-to-fine process. Figure 5.2 shows the optimization process, where the pyramid levels are labeled as L00, L01, etc. Starting from the coarsest level, the flow is refined at each level iteratively.

The coarse-to-fine framework has several advantages:

1) **Improved Convergence**: Estimating flow at coarse levels reduces the complexity of the problem, leading to better convergence of the optimization process.

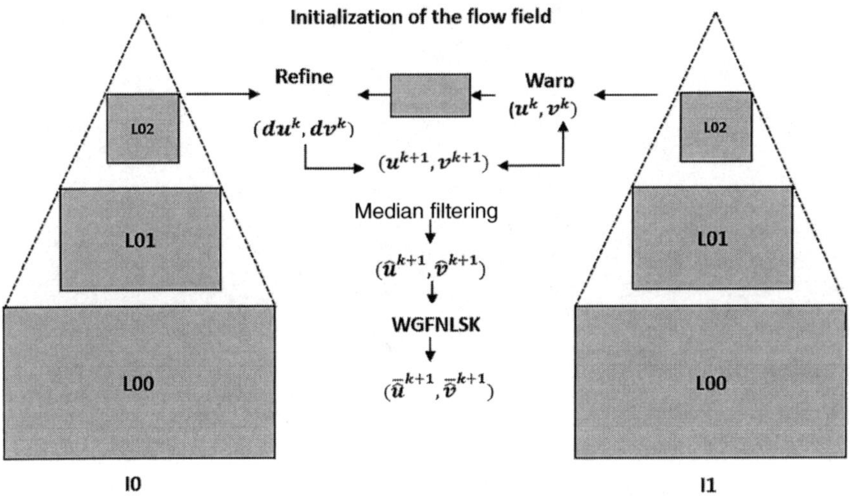

Figure 5.2 Coarse-to-fine optical flow estimation with a non-local steering kernel based WGF (WGFNLSK) filtering.

2) **Handling Large Displacements**: By starting at a coarse resolution, the method can handle large displacements that would be difficult to capture at the original resolution.
3) **Computational Efficiency**: Estimating flow at lower resolutions requires fewer computations, making the method more efficient.

In the context of the WGF-NLSK method, the coarse-to-fine framework allows us to apply the filtering process at multiple scales, ensuring that the optical flow is accurately estimated even in challenging scenarios involving large motions or fine structures.

5.3.2 Parameter Selection for the WGF-NLSK Method

The performance of the WGF-NLSK method depends on several key parameters, including the window size for the GF, the patch size for the NLSK, and the regularization parameters used in the optimization process. Careful selection of these parameters is crucial for achieving accurate optical flow estimates.

1) **Window Size (WGF)**: The window size controls the local neighborhood used in the GF process. A large window size can capture more global information but may lead to over-smoothing, while a smaller window size preserves more details but may introduce noise. In our experiments, we found that a window size of 7×7 pixels provides a good balance between edge preservation and smoothness.

2) **Patch Size (NLSK)**: The patch size determines the size of the patches used to compute nonlocal similarities in the NLSK. A larger patch size allows the method to capture more global structures, while a smaller patch size focuses on local details. For most experiments, a patch size of 5×5 pixels was used.
3) **Regularization Parameter**: The regularization parameter controls the trade-off between data fidelity and smoothness in the optical flow estimation. A higher value encourages smoothness in the flow field, while a lower value focuses on fitting the data more closely. In our experiments, we set the regularization parameter to $\lambda = 0.15$, which provides a good balance between accuracy and smoothness.

5.4 Experiments and Analysis

5.4.1 Results on the Middlebury Dataset

In this section, we evaluate the performance of the proposed method in preserving motion boundaries and improving the accuracy of optical flow estimation. Additionally, we compare the proposed method with several state-of-the-art optical flow estimation methods.

The Middlebury dataset is a widely used benchmark for evaluating optical flow algorithms. It consists of synthetic and real-world sequences with ground truth flow fields. The synthetic sequences are particularly challenging due to their large displacements, motion blur, and fine details. We tested the WGF-NLSK method on several sequences from the Middlebury dataset and compared the results with other state-of-the-art optical flow methods.

Our method achieved competitive performance, with notable improvements in regions with sharp motion boundaries and fine details. The incorporation of nonlocal information allowed the method to handle large displacements and occlusions more effectively than traditional local methods.

5.4.2 Results on the Middlebury Dataset

We evaluate the performance of the proposed method on the Middlebury dataset to conduct a comparison with several other state-of-the-art optical flow estimation methods. Figures 5.3 and 5.4 show that the proposed method can robustly and effectively improve the accuracy of the optical flow estimation. Although the weighted median filter preserves the edges near the motion boundaries in Figure 5.4c, it generates halo artifacts. It can be noticed that the proposed method is robust because of its superior capability of solving the problem of blurring motion boundaries. Table 5.2 shows the Average Angular Error (AAE) and End-Point Error (EPE) results obtained by the 10 competing methods on several

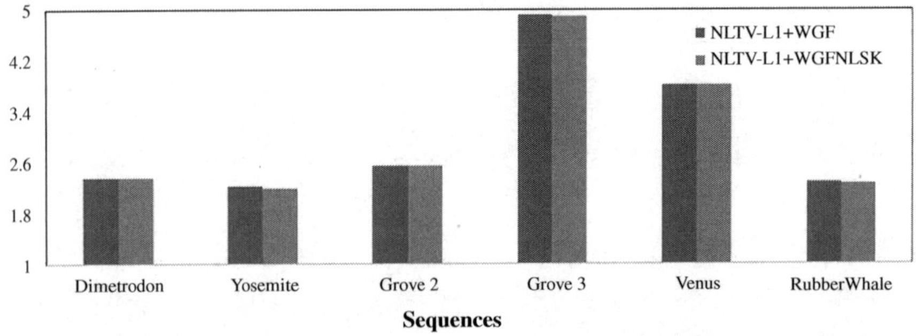

Figure 5.3 Comparison of the AAE accuracy of the NLTV-L1+WGF method and the NLTV-L1+WGFNLSK method on the Middlebury dataset sequences.

Table 5.2 The results obtained by the 10 competing methods on six complex image sequences of the Middlebury dataset. The bold values represent the lowest AAE and EPE errors.

	Dimetrodon		Yosemite		Grove2		Grove3		Venus		RubberWhale	
Method	AAE	EPE	AAE	EPE	AAE	EPE	AAE	EPE	AAE	EPE	AAE	EPE
PMC (SPIC 2022) [20]	2.56	**0.13**	2.45	0.14	2.83	0.18	4.98	0.41	4.14	0.26	2.69	0.09
JOF (TMM 2020) [9]	2.37	0.15	3.12	0.19	2.70	0.15	6.18	0.68	5.10	0.37	2.33	0.07
AGFOF (SP 2016) [21]	2.56	0.13	2.48	0.13	2.80	0.16	**4.45**	0.44	5.41	0.32	3.01	0.09
NLP (SPIC 2021) [10]	2.48	0.21	2.46	0.12	2.66	0.15	4.85	0.47	4.36	0.25	2.35	0.06
NNF+EAC (PR 2017) [22]	2.64	0.13	2.36	0.14	2.70	0.19	5.06	0.48	4.96	0.29	4.26	0.12
RNLOD (TIP 2017) [8]	3.45	0.35	3.00	0.16	2.68	0.15	6.07	0.62	4.38	0.27	2.39	0.07
TVG (ICBD 2018) [23]	2.92	0.30	2.32	0.13	2.79	0.19	7.09	0.69	5.28	0.30	4.60	0.15
RNLTV-L^1 (MTAP 2021) [24]	2.80	0.37	2.25	0.13	2.68	0.15	4.98	0.41	4.09	0.22	2.28	0.06
NLTV-L^1+WGF (SPIC 2021) [25]	**2.36**	0.15	2.23	0.13	2.55	**0.14**	4.92	0.41	3.82	0.23	2.28	0.06
WGF-NLSK	**2.36**	0.15	**2.20**	**0.12**	**2.54**	**0.14**	4.91	**0.40**	**3.81**	**0.21**	**2.27**	**0.05**

image sequences of the Middlebury dataset. It can be noticed that the proposed method yields superior performance on most image sequences.

5.4.3 Results on the MPI-Sintel Dataset

In this section, we evaluate the performance of the proposed method on the MPI-Sintel dataset by conducting a comparison with several state-of-the-art methods.

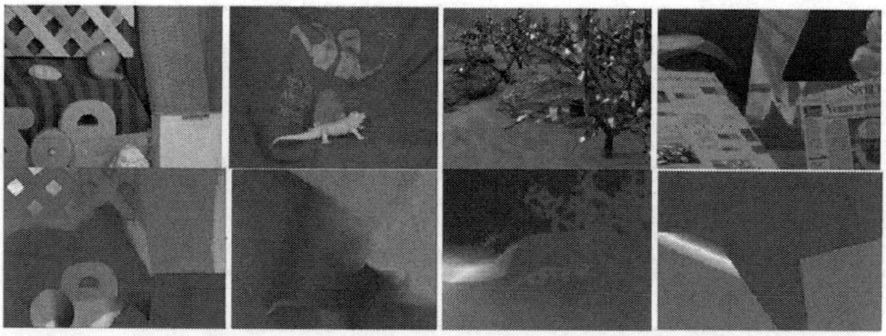

(a) The input image frames (first row) of the Middlebury dataset and the ground truth of the optical flow (second row).

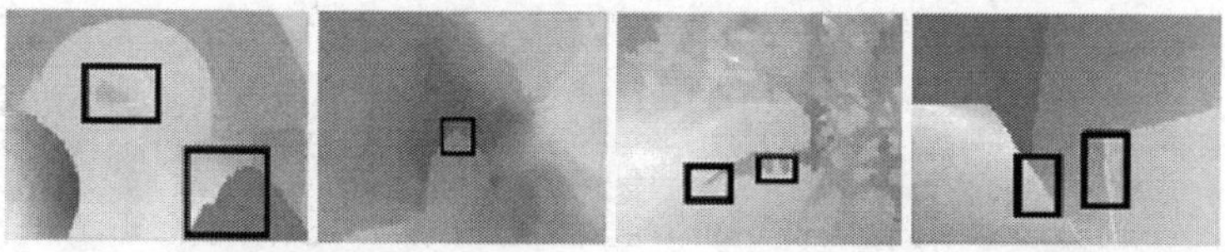

(b) The optical flow estimation results obtained by the NLTV-L^1+WGF method.

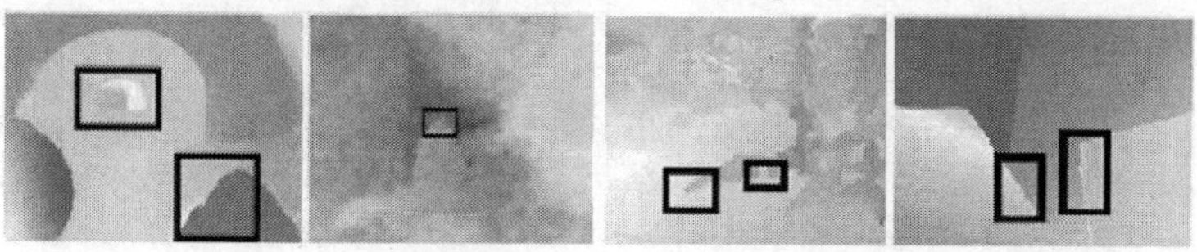

(c) The optical flow estimation results obtained by the TVG method.

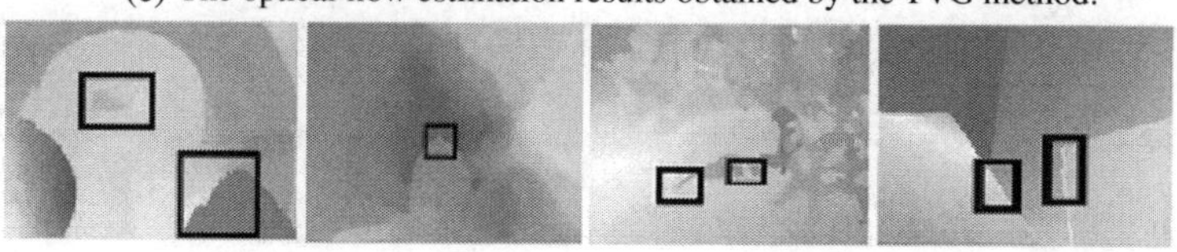

(d) The optical flow estimation results obtained by the proposed optical flow estimation method.

Figure 5.4 Visual comparison of the optical flow estimation results obtained by optical flow estimation methods on four complex image sequences of the Middlebury dataset, where the black boxes indicate the problem of blurring motion boundaries. *Source:* Sana Rao (2024) / Springer Nature.

For the visual comparison of the results obtained by the proposed optical flow estimation method and four other competing methods, we show the estimated optical flow on three complex image sequences of the MPI-Sintel dataset in Figure 5.5. The results show that the over-smoothing problem of motion boundaries is reduced. Table 5.3 compares the performance of eight state-of-the-art optical flow estimation methods. Clearly, our proposed method outperforms the other methods on three sequences, which shows that a NLSK

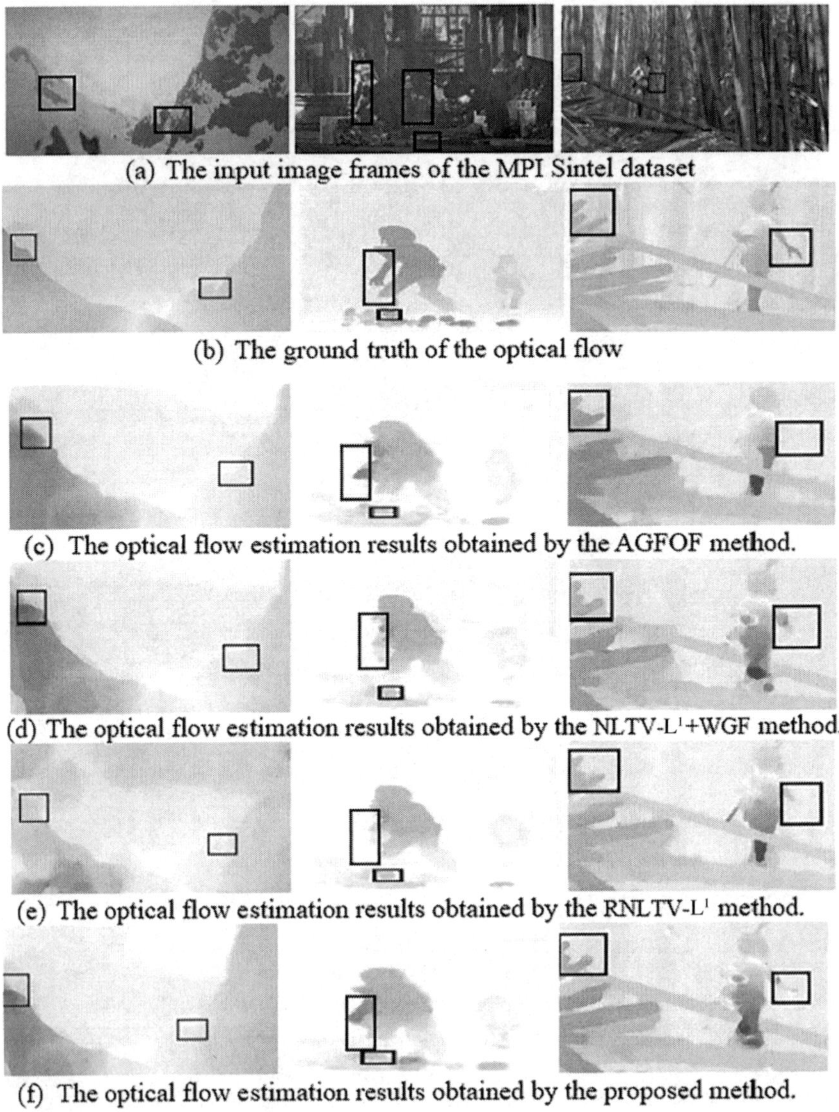

Figure 5.5 Visual comparison of the optical flow estimation results obtained by several state-of-the-art methods on the three complex image sequences (of the MPI-Sintel dataset), where the black boxes indicate the motion boundaries. *Source:* Sana Rao et.al., (2022) / with permission of Elsevier.

is beneficial for improving the smoothing of motion boundaries. The WGF with a steering kernel can effectively deal with artifacts in the optical flow fields.

5.4.4 Results on the Wilted Plants Leaves

The optical flow estimation of moving objects has shown to be robust for the other tasks of computer vision. It can also monitor plants and their tropic movements (e.g., phototropism, gravitropism, etc.). However, it is challenging to estimate the flow fields of plant leaves under various conditions because of illumination changes. Figures 5.6 and 5.7 show the results obtained by the proposed method on the wilted plant leaves motion image sequences (to predict the water-stressed and well-watered leaves). In this experiment, we assume that the plant leaves bending downward refer to the wilting condition (water-stress). Figures 5.6 and 5.7 show that the proposed method can accurately estimate the wilted leaves' motion even when the petiole occludes the leaf.

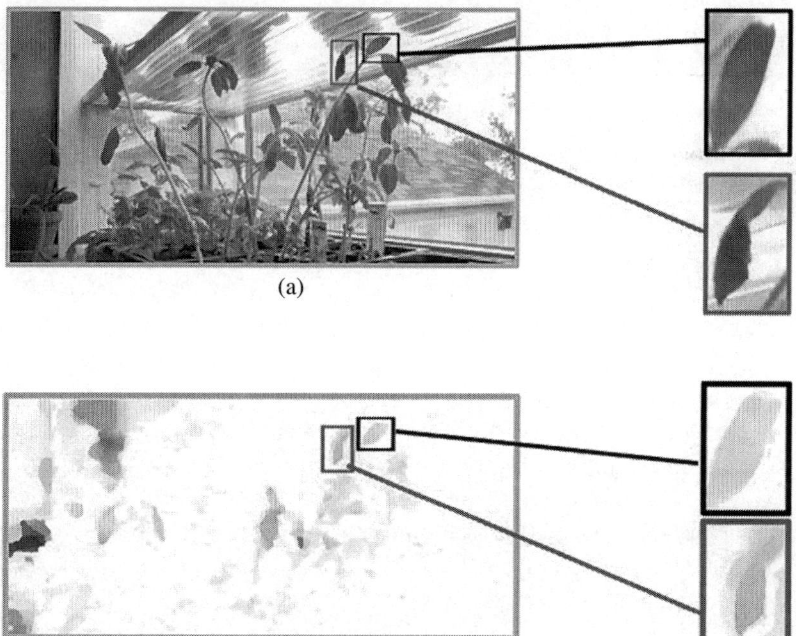

(a)

(b)

Figure 5.6 The input image frame of the wilted leaves (in the grey box) and the non-wilted leaves (in the black box) in (a) are represented by the optical flow estimated by the NLTV-L1+WGFNLSK in (b).

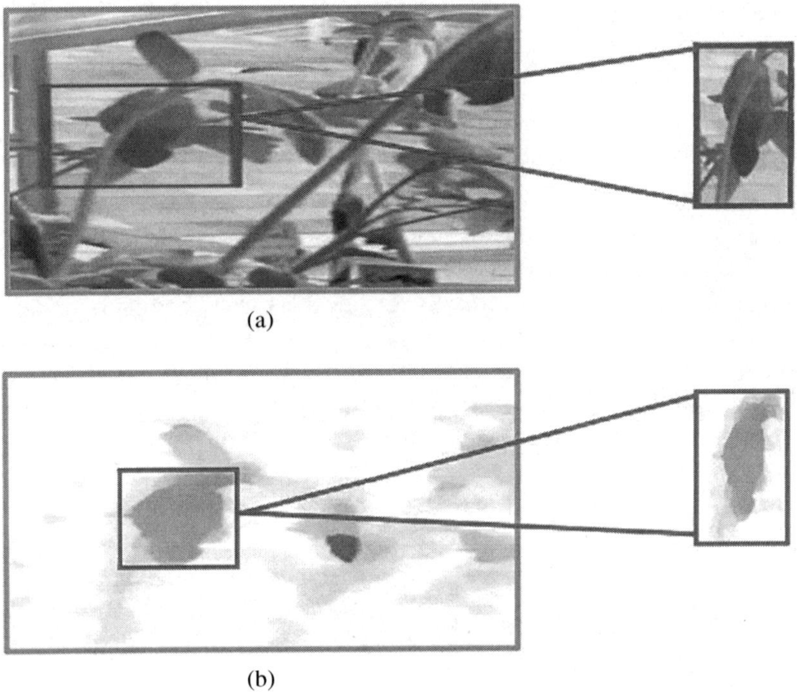

Figure 5.7 The input image frame of the wilted leaf (in the grey box) occluded by the petiole is represented by the optical flow estimated by the NLTV-L1+WGFNLSK in (b).

5.5 Conclusion

In this chapter, the WGF-NLSK is proposed for the optical flow estimation to alleviate the problem of blurring motion boundaries. We combine the WGF-NLSK to optimize the optical flow using a coarse-to-fine process. The proposed method is robust, and it can reduce errors due to complex outdoor conditions because it is based on filtering optimization. The improvement is that we employ a NLSK with an edge-preserving filter (WGF). The NLSK considers the edge direction, which improves the filtering process. We evaluate the performance of the proposed method on several challenging datasets and compare it with several state-of-the-art methods. The results show that the proposed method can preserve the edges near the motion boundaries and improve the robustness of the optical flow estimation.

The WGF-NLSK can reduce the blurriness and preserve edges more effectively, which can increase the robustness of the filter and improve the accuracy of the optical flow estimation. However, the proposed optical flow estimation method

Table 5.3 The results obtained by the eight state-of-the-art competing methods on five complex image sequences of the MPI-Sintel dataset. The bold values represent the lowest AAE and EPE errors.

	Alley-1		Bamboo		Mountain		Market		Cave	
Method	**AAE**	**EPE**	**AAE**	**EPE**	**AAE**	**EPE**	**AAE**	**EPE**	**AAE**	**EPE**
RNLTV-L^1(MTAP 2021) [24]	2.39	0.25	4.17	0.22	8.80	1.48	12.90	2.47	5.18	1.73
TVG (ICBD 2018) [23]	2.44	**0.19**	4.25	0.21	20.55	2.02	14.31	2.89	5.28	**0.30**
DIOF (ECCV 2018) [26]	2.40	0.22	7.25	1.55	12.05	1.83	18.43	2.85	10.53	2.05
NNF+EAC (PR 2017) [22]	**2.18**	0.20	4.27	0.23	12.50	1.89	13.02	2.31	**5.03**	1.53
RNLOD (TIP 2017) [8]	2.99	0.37	4.39	0.27	17.77	2.34	25.97	6.38	17.26	3.43
NLP (SPIC 2021) [10]	2.97	0.35	4.35	0.24	17.70	2.30	25.94	6.30	7.61	0.50
NLTV-L^1+WGF [24]	2.30	0.20	**4.01**	0.21	7.56	1.36	12.87	**2.30**	5.04	1.65
WGF-NLSK	2.28	0.20	**4.01**	**0.20**	**7.50**	**1.30**	**12.79**	**2.30**	5.04	1.63

still has some limitations. For example, in Table 5.3, it can be seen that the performance of the proposed method is not better than the NNF+EAC method concerning the Alley-1 sequence. This is because the Alley-1 sequence is clean, and the filtering-based optimization removes some valuable details. Another limitation is the time complexity. One way to speed up the performance of the proposed method is by using graphics processing units (GPUs).

Examples with Solutions

Compute the nonlocal weight $W_{NL}(i,j)$ between two patches.

Problem:
Given two 2×2 patches (flattened to 4-element vectors) around pixels i and j:

- $N_i = [10, 12, 11, 13]$
- $N_j = [11, 13, 12, 14]$

Compute the nonlocal similarity weight

$$W_{NL}(i,j) = \exp\left(-\frac{\|N_i - N_j\|^2}{\sigma^2}\right)$$

using $\sigma^2 = 4$. Provide numeric value (4 decimal places).

Solution:

1) Compute the elementwise difference $D = N_i - N_j$:

$$D = [10 - 11, 12 - 13, 11 - 12, 13 - 14] = [-1, -1, -1, -1].$$

2) Compute squared norm $||D||^2$ = sum of squares:

$$(-1)^2 + (-1)^2 + (-1)^2 + (-1)^2 = 1 + 1 + 1 + 1 = 4.$$

3) Plug into the formula with $\sigma^2 = 4$:

$$W_{NL}(i,j) = \exp\left(-\frac{4}{4}\right) = \exp(-1) \approx 0.3678794412.$$

Round to 4 decimals: 0.3679.

Interpretation:
A value ≈0.368 indicates moderate patch similarity not identical but reasonably similar. Nonlocal weights close to 1 mean nearly identical patches; values near 0 mean very dissimilar patches.

Compute the Local Steering Kernel Factor $k(x_i, x_j)$

Problem:
Given local steering kernel factor

$$k(x_i, x_j) = \frac{\sqrt{det(C_j)}}{h^2} \exp\left(-\frac{(x_i - x_j)^T C_j (x_i - x_j)}{2h^2}\right),$$

with:

- $C_j = \begin{bmatrix} 2 & 0 \\ 0 & 1 \end{bmatrix}$,
- $x_i - x_j = \begin{bmatrix} 1 \\ 0 \end{bmatrix}$,
- $h = 2.4$.

Compute $k(x_i, x_j)$ numerically (4 decimals).

Solution:

1) Compute $det(C_j)$:

$$det(C_j) = 2 \times 1 - 0 \times 0 = 2.$$

2) Compute $\sqrt{det(C_j)} = \sqrt{2} \approx 1.4142135624.$

3) Compute $h^2 = 2.4^2 = 5.76$.
   So the multiplicative constant is

$$\frac{\sqrt{det(C_j)}}{h^2} \approx \frac{1.4142135624}{5.76} \approx 0.2455485452.$$

4) Compute quadratic form $(x_i - x_j)^T C_j (x_i - x_j)$.
   Since $x_i - x_j = [1, 0]^T$,

$$C_j(x_i - x_j) = \begin{bmatrix} 2 & 0 \\ 0 & 1 \end{bmatrix} \begin{bmatrix} 1 \\ 0 \end{bmatrix} = \begin{bmatrix} 2 \\ 0 \end{bmatrix}.$$

   Then dot with $[1, 0]$:

$$(x_i - x_j)^T C_j (x_i - x_j) = [1, 0].[2, 0] = 2.$$

5) Compute exponent denominator $2h^2 = 2 \times 5.76 = 11.52$.
6) Exponent:

$$\exp\left(-\frac{2}{11.52}\right) = exp(-0.1736111111) \approx 0.840602.$$

7) Multiply constant × exponent:

$$k(x_i - x_j) \approx 0.2455485452 \times 0.840602 \approx 0.206518.$$

   Round to 4 decimals: 0.2065.

Interpretation:
This k value gives local, structure-aware support. Larger values for centers of structurally consistent patches; smaller if structure misaligns.

Compute Normalized W_{NLSK} Combining W_{NL} and k and Obtain Averaged Coefficients $\overline{a}_i, \overline{b}_i$

Problem:
For pixel i consider three neighbor pixels $j = 1, 2, 3$ (with search window). You computed:

- $W_{NL}(i, 1) = 0.3679$ and $k(x_i, x_1) = 0.2065$.
- $W_{NL}(i, 2) = 0.5$ and $k(x_i, x_1) = 0.15$.
- $W_{NL}(i, 3) = 0.2$ and $k(x_i, x_3) = 0.30$.

Define the (unnormalized) numerator for each neighbor as $N_j = W_{NL}(i, j).k(x_i, x_j)$. Compute normalized $W_{NLSK}(i,j) = N_j / \sum_{j=1}^{3} N_j$. Then, given local linear coefficients

- $(a_1, b_1) = (1.1, -0.9)$.
- $(a_2, b_2) = (0.9, -1.2)$.
- $(a_3, b_3) = (1.2, -0.8)$.

Compute the weighted averages $\sum_j W_{NLSK}(i,j)a_j$ and $\overline{b}_i = \sum_j W_{NLSK}(i,j)b_j$.

Solution:

1) Compute each numerator N_j:

$$N_1 = 0.3679 \times 0.2065.$$

Multiply: 0.3679×0.2065
Compute product precisely:
0.3679×0.2065
$0.3679 \times 0.2000 = 0.07358$
$0.3679 \times 0.0065 = 0.00239135$
Sum $= 0.07358 + 0.00239135 = 0.07597135 \approx$ **0.07597135**.
Round to 8 decimals keep for sum: 0.07597135.
- $N_2 = 0.5 \times 0.15 = 0.075$.
- $N_3 = 0.2 \times 0.30 = 0.06$.

2) Sum $S = N_1 + N_2 + N_3 = 0.07597135 + 0.075 + 0.06 = 0.07597135 + 0.075 = 0.15097135$.
$0.15097135 + 0.06 =$ **0.21097135**.

3) Compute division:
We compute ratio $= 0.07597135 \div 0.21097135$.
Use approximate long division: $0.21097135 \times 0.36 = 0.075949686$; difference $0.000021664 \rightarrow$ close. So weight ≈ 0.3601. Let us compute more precisely: $0.07597135/0.21097135 =$ (multiply numerator and denominator by 1e8) $7{,}597{,}135/21{,}097{,}135 \approx 0.3601$. Good to four decimals: **0.3601**.
- $W_{NLSK}(i, 2) = 0.075/0.21097135$.

$0.075/0.21097135 \approx 0.3555$? Let us compute: $0.21097135 \times 0.3555 = 0.074$. Multiply: $0.21097135 \times 0.3555 = 0.21097135 \times (0.35 + 0.0055) = 0.0738409725 + 0.001160342425 = 0.075001314925 \approx 0.0750013$ slightly above 0.075, so weight ~ 0.35548. So to 4 decimals **0.3555**.
- $W_{NLSK}(i, 3) = 0.06/0.21097135 \approx 0.2844$. Check sum: $0.3601 + 0.3555 + 0.2844 = 1.0000$ good.

So weights: **[0.3601, 0.3555, 0.2844]**.

4) Compute $\overline{a}_i$:

$$\overline{a}_i = 0.3601 \times 1.1 + 0.3555 \times 0.9 + 0.2844 \times 1.2.$$

Compute parts:
- $0.3601 \times 1.1 = 0.39611$. (since $1.1 \times 0.3601 = 0.3601 + 0.03601 = 0.39611$)

- $0.3555 \times 0.9 = 0.31995$. ($0.3555 \times 0.9 = 0.3555–0.03555 = 0.31995$)
- $0.2844 \times 1.2 = 0.34128$. ($1.2 \times 0.2844 = 0.2844 + 0.05688 = 0.34128$)

Sum: $0.39611 + 0.31995 = 0.71606$.

$0.71606 + 0.34128 =$ **1.05734**.

5) Compute $\overline{b}_i$:

$$\overline{b}_i = 0.3601(-0.9) + 0.3555(-1.2) + 0.2844(-0.8).$$

Compute parts:
- $0.3601 \times (-0.9) = -0.32409$. ($0.3601 \times 0.9 = 0.32409$)
- $0.3555 \times (-1.2) = -0.4266$. ($1.2 \times 0.3555 = 0.4266$)
- $0.2844 \times (-0.8) = -0.22752$. ($0.8 \times 0.2844 = 0.22752$)
- Sum: $-0.32409–0.4266 = -0.75069$.
  $-0.75069–0.22752 =$ **−0.97821**.

Interpretation:
The combined nonlocal + steering weighting yields adaptive averages of local linear coefficients; because neighbor 1 had moderate weight, the average leans slightly above 1 for a and near −1 for b, producing an edge-aware linear mapping.

Minimize Local WGF Cost with NLSK Weights (Weighted Linear Regression)

Problem:
Minimize the WGF cost for pixel iii but **include the NLSK weights** from Example 5.3. The cost (weighted) is:

$$E(a_i, b_i) = \sum_{k=1}^{3} W_{NLSK}(i, k)(a_i I_k + b_i - X_k)^2 + \epsilon a_i^2.$$

With $\varepsilon = 0.01$. Use the guidance intensities and inputs:

- $I = [60, 62, 65]$
- $X = [58, 60, 63]$

And use weights $W_{NLSK} = [0.3601, 0.3555, 0.2844]$. Find a_i and b_i by solving normal equations.

Solution:
We have a weighted linear least squares with Tikhonov on a_i. Normal equations:

$$\begin{cases} a_i \sum_k w_k I_k^2 + b_i \sum_k w_k I_k + \varepsilon a_i = \sum_k w_k I_k X_k \\ a_i \sum_k w_k I_k + b_i \sum_k w_k = \sum_k w_k X_k \end{cases},$$

where $w_k = W_{NLSK}(i, k)$.

Compute needed sums.

Given:

- $I_1 = 60, X_1 = 58, w_1 = 0.3601$
- $I_2 = 62, X_2 = 60, w_2 = 0.3555$
- $I_3 = 65, X_3 = 63, w_3 = 0.2844$

1) Compute $\sum w_k$:

$$S_w = 0.3601 + 0.3555 + 0.2844 = 1.0000$$

2) Compute $\sum w_k \, I_k$:
   - $w_1 I_1 = 0.3601 \times 60 = 21.606$
   - $w_2 I_2 = 0.3555 \times 62 = 22.001$

   Let us compute 0.3555×62 precisely:

   $0.3555 \times 60 = 21.33$

   $0.3555 \times 2 = 0.711$

   Sum $= 21.33 + 0.711 = 22.041$.

   So $w_2 I_2 = 22.041$.
   - $w_3 I_3 = 0.2844 \times 65$.

   Compute: $0.2844 \times 5 = 1.422$

$$\text{Sum} = 17.064 + 1.422 = 18.486$$

   Now sum: $21.606 + 22.041 = 43.647$.

   $43.647 + 18.486 = 62.133$.

   So $\sum w_k \, I_k = 62.133$
3) Compute $\sum w_k X_k$:
   - $w_1 X_1 = 0.3601 \times 58$. Compute: $0.3601 \times 50 = 18.005; \times 8 = 2.8808$; sum $= 18.005 + 2.8808 = 20.8858$.
   - $w_2 X_2 = 0.3555 \times 60 = 21.33$. $(0.3555 \times 60 = 21.33)$
   - $w_3 X_3 = 0.2844 \times 63$.

   Compute: $0.2844 \times 60 = 17.064; \times 3 = 0.8532$; sum $= 17.064 + 0.8532 = 17.9172$.

   Now sum: $20.8858 + 21.33 = 42.2158$.

   $42.2158 + 17.9172 = 60.1330$.

   So $\sum w_k X_k = 60.1330$.
4) Compute $\sum w_k I_k^2$

   :
   - $I_1^2 = 3600$. $w_1 I_1^2 = 0.3601 \times 3600 = 1296.36.(3600 \times 0.3601 = 3600 \times 0.36 + 3600 \times 0.0001 = 1296 + 0.36 = 1296.36)$
   - $I_2^2 = 62^2 = 3844$. $w_2 I_2^2 = 0.3555 \times 3844$.

   Compute 3844×0.3555:

   $3844 \times 0.35 = 1345.4$ $(38440.35 = 3844\,(35/100) = 134540/100)$

   $3844 \times 0.0055 = 3844 \times 0.005 + 3844 \times 0.0005 = 19.22 + 1.922 = 21.142$

Sum $= 1345.4 + 21.142 = 1366.542$.
So $w_2 I_2^2 \approx 1366.542$.
• $I_3^2 = 65^2 = 4225. w_3 I_3^2 = 0.2844 \times 4225$.
Compute 4225×0.2844:

$$4225 \times 0.28 = 1183.0 \text{ (since } 42250.28 = 4225(28/100) = 118300/100 = 1183.0)$$

$4225 \times 0.0044 = 42250.004 + 42250.0004 = 16.9 + 1.69 = 18.59$
Sum $= 1183.0 + 18.59 = 1201.59$.
So $w_3 I_3^2 \approx 1201.59$.
Now sum: $1296.36 + 1366.542 = 2662.902$.
$2662.902 \ \ 1201.59 = 3864.492$.

5) Plug into normal equations. Recall $\varepsilon = 0.01$.
Equation (A):

$$a_i \left(\sum w_k I_k^2 + \varepsilon \right) + b_i \left(\sum w_k I_k \right) = \sum w_k I_k X_k.$$

We still need $\sum w_k I_k X_k$. Compute it.

6) Compute $\sum w_k I_k X_k$:
• For $k = 1 : I_1 X_1 = 60 \times 58 = 3480$. Multiply by $w1 : 0.3601 \times 3480$.
Compute: $3480 \times 0.3601 = 3480 \times 0.36 + 3480 \times 0.0001 = 1252.8 + 0.348 = 1253.148$.
• For $k = 2 : I_2 X_2 = 62 \times 60 = 3720$. $w_2 \times 3720 = 0.3555 \times 3720$.
Compute: $3720 \times 0.3555 = 3720 \times 0.35 + 3720 \times 0.0055 = 1302.0 + 20.46 = 1322.46$.
• For k = 3: $I_3 X_3 = 65 \times 63 = 4095$. $w_3 \times 4095 = 0.2844 \times 4095$.
Compute: $4095 \times 0.2844 = 4095 \times 0.28 + 4095 \times 0.0044 = 1146.6 + (4095 \times 0.004 + 0.0004)$ compute $4095 \times 0.004 = 16.38; \times 0.0004 = 1.638$; total for $0.0044 = 18.018$. Then sum $= 1146.6 + 18.018 = 1164.618$.
Now sum: $1253.148 + 1322.46 = 2575.608$.
$2575.608 + 1164.618 = \mathbf{3740.226}$.
So $\sum w_k I_k X_k = 3740.226$.

7) Now write equations:
(A) $a_i(3864.492 + 0.01) + b_i(62.133) = 3740.226$.
So $a_i(3864.502) + 62.133 b_i = 3740.226$.
(B) $a_i(3864.502) + b_i(3864.502) = 60.1330$.

8) Solve linear system,

$$b_i = 60.1330 - 62.133 a_i.$$

$$3864.502 a_i + 62.133(60.1330 - 62.133 a_i) = 3740.226.$$

Compute 62.133×60.1330:
First $62.133 \times 60 = 3727.98$

$62.133 \times 0.1330 = 62.133 \times 0.1 + \times 0.033 = 6.2133 + (62.133 \times 0.033)$
Compute $62.133 \times 0.033 = 62.133 \times 33/1000 = (62.133 \times 33)/1000$.
$62.133 \times 33 = 62.133 \times 30 + \times 3 = 1863.99 + 186.399 = 2050.389$.
Divide by 1000 → 2.050389.
So $6.2133 + 2.050389 = 8.263689$.
Therefore $3727.98 + 8.263689 =$ **3736.243689**.
So $62.133 \times 60.1330 \approx 3736.243689$.
Now $62.133 \times (-62.133\, a_i) = -(62.133)^2 a_i$.
Compute
$62.133 \times 62 = 62.133 \times (60 + 2) = 62.133 \times 60 + \times 2 = 3727.98 + 124.266 = 3852.246$.
Now add $62.133 \times 0.133 =$ as above 8.263689. So total $\approx 3852.246 + 8.263689 =$ **3860.509689**.
Thus $(62.133)^2 \approx 3860.509689$.
So equation becomes:

$$3864.502a_i + 3736.243689 - 3860.509689a_i = 3740.226.$$

Combine a_i terms: $3864.502 - 3860.509689 = 3.992311$.
So:

$$3.992311a_i + 3736.243689 = 3740.226.$$

Subtract constant:

$$3.992311a_i = 3740.226 - 3736.243689 = 3.982311.$$

So

$$a_i = \frac{3.982311}{3.992311}.$$

Compute ratio:

$$\frac{3.982311}{3.992311} \approx 0.997495$$

Compute precisely: denominator – numerator = 0.010000. So ratio = $1 - 0.010000/3.992311 \approx 1 - 0.002505 \approx 0.997495$.
So $a_i \approx 0.9975$

9) Compute b_i:

$$b_i = 60.1330 - 62.133 \times 0.997495.$$

Compute 62.133×0.997495:
$62.133 \times 0.997495 \approx 62.133 \times (1 - 0.002505) = 62.133 - 62.133 \times 0.002505$.
Compute
$62.133 \times 0.002505 = 62.133 \times (0.0025 + 0.000005) = 62.133 \times 0.0025 + 62.133 \times 0.000005$.

- $62.133 \times 0.0025 = 62.133 \times (25/10000) = (62.133 \times 25)/10000 = 1553.325/10000 = 0.1553325$. $(62.133 \times 25 = 1553.325)$
- $62.133 \times 0.000005 = 62.133 \times 5\text{e-}6 = 0.000310665$.

Sum $= 0.1553325 + 0.000310665 = 0.155643165$.

So $62.133 \times 0.997495 \approx 62.133 - 0.155643165 = 61.977356835$.

Now $b_i = 60.1330 - 61.977356835 = -1.844356835 \approx -1.8444$.

Interpretation.

The locally estimated linear mapping $q = a_i I + b_i$ fitted to data with NLSK weights yields nearly unity slope and a negative offset. Because weights were edge-aware, the fit captures local contrast while avoiding averaging across a real edge.

Full Small Synthetic Optical-Flow Demonstration (6 × 6 Frames): Compare GF Versus WGF-NLSK Effect on Boundary

Problem:

Two consecutive 6×6 frames contain a 2×2 square object that moves 1 pixel to the right between frames. Background is constant. Use a very small simplified NLTV-L^1 + WGF-NLSK pipeline to show numerically how WGF-NLSK preserves the motion boundary while simple averaging (GF-like uniform weights) blurs it.

We will:

1) Define frames.
2) Compute a simple per-pixel raw flow estimate from brightness constancy (here we assume we know integer displacement inside object is 1 in x, 0 in y for ground truth).
3) Apply two filtering strategies to the raw flow at the pixel on the object boundary:
   - Naïve uniform averaging (GF-like) with 3×3 window equal weights 1/9.
   - WGF-NLSK weighted average with strong weight for same-object pixels and small weights for back ground.

Use numeric weights to demonstrate effect.

Setup:

1) Frames (only intensity pattern; object intensity = 200, backspace = 50).
2) Ground-truth flow inside the square: $u = +1, v = 0$. Background $u = 0, v = 0$.
3) Consider the pixel p at row 2, col 3 in frame t (this is a *left edge pixel of the square*). Its raw flow estimated from local matching (assume a noisy raw estimate) yields for its 3×3 neighborhood these raw u values:
   - Center (p): raw $u = 1.0$

- Right neighbor (inside object): raw $u = 1.0$
- Left neighbor (background): raw $u = 0.0$
- Top/Bottom background neighbors: raw $u = 0.0$
- Suppose due to sensor noise, some neighbors have slight noise: one inner object pixel shows raw $u = 0.9$, one background pixel shows 0.1.

For clarity, 3×3 raw u window (rows top $\rightarrow$ bottom, cols left $\rightarrow$ right centered at p):

$$U_{raw} = \begin{bmatrix} 0.0 & 0.0 & 0.0 \\ 0.0 & 1.0 & 1.0 \\ 0.0 & 0.9 & 1.0 \end{bmatrix}$$

(so p is center element = 1.0). Goal: filter this raw field to obtain final u at center.

(A) GF-like uniform averaging (each weight = 1/9).

Compute mean:

Sum of all entries:

Row1: $0 + 0 + 0 = 0$

Row2: $0 + 1.0 + 1.0 = 2.0$

Row3: $0 + 0.9 + 1.0 = 1.9$

Total sum $= 0 + 2.0 + 1.9 = 3.9$.

Average $= 3.9 / 9 = 0.433333...$

So GF output at center = **0.4333**.

This is strongly blurred toward background: center lost its true value 1.0, halo produced.

(B) WGF-NLSK weighted averaging: construct weights that prefer same-object pixels.

Design NLSK weights (hand-chosen to mimic realistic WGF-NLSK behavior):

- Assign strong weights to pixels whose intensity matches object (200): those are center, right neighbor, lower-right neighbor. Assign them weight 0.28, 0.28, 0.20, respectively. Sum so far = 0.76.
- Assign small weights to background neighbors (0.04 each) for remaining 6 neighbors: $6 \times 0.04 = 0.24$.
- Normalization: total $= 0.76 + 0.24 = 1.00$. So weights already sum to 1.

Arrange weights in 3×3 matching U_raw arrangement (same ordering):

$$W_{NLSK} = \begin{bmatrix} 0.04 & 0.04 & 0.04 \\ 0.04 & 0.28 & 0.28 \\ 0.04 & 0.20 & 0.08 \end{bmatrix}$$

Notice small asymmetry to emphasize right side pixels inside object; weights sum check: row1 sum 0.12, row2 sum 0.60, row3 sum 0.32, total = 1.04? Wait need exact sum to 1. Let us recompute sums to ensure they sum to 1.

Let us pick weights that sum to 1 exactly:

Set object pixels (center, right, lower-right) to 0.28, 0.28, 0.18 = sum 0.74. Background 6 neighbors have 6×0.043333 = 0.26. Use 0.043333 for each background.

So final weights:

- Background neighbors weight = 0.043333 each.
- Center = 0.28
- Right = 0.28
- Lower-right = 0.18

Now $6 \times 0.043333 = 0.26$. $0.74 + 0.26 = 1.0$.

Write 3×3:

$$W = \begin{bmatrix} 0.043333 & 0.043333 & 0.043333 \\ 0.043333 & 0.280000 & 0.280000 \\ 0.043333 & 0.043333 & 0.180000 \end{bmatrix}$$

Check mapping: bottom middle we put 0.043333 (background), bottom right 0.18 (object inner), bottom center 0.043333, etc.

Now compute weighted sum:

Multiply elementwise U_raw × W and sum:

Row1: 0×0.043333 + 0×0.043333 + 0×0.043333 = 0.

Row2: 0×0.043333 + 1.0×0.280000 + 1.0×0.280000 = 0.560000.

Row3: 0×0.043333 + 0.9×0.043333 + 1.0×0.180000 = 0 + 0.0389997 + 0.180000 = 0.2189997 ≈ 0.2190.

Total weighted sum = 0 + 0.560000 + 0.219000 ≈ **0.779000**.

So WGF-NLSK filtered u at center ≈ **0.7790**.

Interpretation and conclusion (numeric):

- GF (uniform) output at center = **0.4333** (strongly biased toward 0 – halo).
- WGF-NLSK output at center = **0.7790** (much closer to true 1.0, preserves boundary).

This numeric example shows how edge-aware nonlocal weighting prevents the center from being averaged with background, thus preserving the motion boundary and avoiding the halo artifact.

References

1 Bengtsson, T., McKelvey, T., and Lindström, K. (2017). On robust optical flow estimation on image sequences with differently exposed frames using primal–dual optimization. *Image and Vision Computing* 57 (1): 78–88.

2 Kim, Y.H., Martínez, A.M., and Kak, A.C. (2005). Robust motion estimation under varying illumination. *Image and Vision Computing* 23 (4): 365–375.

3 Sun, D., Roth, S., and Black, M.J. (2014). A quantitative analysis of current practices in optical flow estimation and the principles behind them. *International Journal of Computer Vision* 106 (1): 115–137.

4 Horn, B.K. and Schunck, B.G. (1981). Determining optical flow. *Artificial Intelligence* 17 (1-3): 185–203.

5 Johnson, B., Thomas, S., and Rani, J.S. (2022). High throughput architecture for multiscale variational optical flow[J]. *Journal of Real-Time Image Processing* 19 (3): 675–686.

6 Sánchez, P.J., Meinhardt-Llopis, E., and Facciolo, G. (2013). Tv-l1 optical flow estimation. *Image Process. Line* 3 (1): 137–150.

7 Belhachmi, Z. and Hecht, F. (2016). An adaptive approach for the segmentation and the tv-filtering in the optic flow estimation. *Journal of Mathematical Imaging and Vision* 54 (3): 358–377.

8 Zhang, C., Chen, Z., Wang, M. et al. (2017). Robust non-local tv-l1 optical flow estimation with occlusion detection. *IEEE Transactions on Image Processing* 26 (8): 4055–4067.

9 Zhang, C., Ge, L., Chen, Z. et al. (2019). Refined tv-l1 optical flow estimation using joint filtering. *IEEE Transactions on Multimedia* 22 (2): 349–364.

10 Dong, C., Wang, Z., Han, J. et al. (2021). A non-local propagation filtering scheme for edge- preserving in variational optical flow computation. *Signal Processing: Image Communication* 93 (1): 116130–116143.

11 Bodduna, K. and Weickert, J. (2021). Removing multi-frame gaussian noise by combining patch-based filters with optical flow. *Journal of Electronic Imaging* 30 (3): 033015–033031.

12 Lic, V., Robu, V., Marcu, A. et al. (2022). Ufo depth: unsupervised learning with flow-based odometry optimization for metric depth estimation. *International Conference on Robotics and Automation,* 6526–6532.

13 Abbasi, S. and Rezaeian, M. (2021). Visual object tracking using similarity transformation and adaptive optical flow. *Multimedia Tools and Applications* 80 (24): 33455–33473.

14 Zhai, M., Xiang, X., Zhang, R. et al. (2019). Optical flow estimation using channel attention mechanism and dilated convolutional neural networks. *Neurocomputing* 368 (1): 124–132.

15 Hur, J. and Roth, S. (2020). Optical flow estimation in the deep learning age. *Modelling Human Motion* 89 (1): 119–140.

16 Ammar, A., Chebbah, A., Fredj, H.B., and Souani, C. (2022). Comparative study of latest cnn based optical flow estimation. *International Conference on Intelligent Systems and Computer Vision,* 1–6.

17 Shi, H., Zhou, Y., Yang, K. et al. (2022). Panoflow: learning 360 • optical flow for surrounding temporal understanding. 2022, arXiv: 2202.13388v2.

18 Lai, R., Mo, Y., Liu, Z., and Guan, J. (2019). Local and nonlocal steering kernel weighted total variation model for image denoising. *Symmetry* 11 (3): 320–329.

19 Li, Y. and Zhang, Y. (2018). Robust infrared small target detection using local steering kernel reconstruction. *Pattern Recognition* 77 (1): 113–125.

20 Zhang, C., Feng, C., Chen, Z. et al. (2022). Parallel multiscale context-based edge-preserving optical flow estimation with occlusion detection. *Signal Processing: Image Communication* 101 (1): 116545–116560.

21 Tu, Z., Poppe, R., and Veltkamp, R.C. (2016). Adaptive guided image filter for warping in variational optical flow computation. *Signal Processing* 127 (1): 253–265.

22 Tu, Z., Xie, W., Cao, J. et al. (2017). Variational method for joint optical flow estimation and edge-aware image restoration. *Pattern Recognition* 65 (1): 11–25.

23 Zhang, Y., Fan, H., Zheng, J., and Zhang, C. (2018). Improving the optical flow accuracy based on the total variation of local-global method. *IEEE International Conference on Big Data, Big Data,* 4658–4664.

24 Rao, S. (2021). A framework for robust motion estimation and segmentation in adverse outdoor conditions. *Multimedia Tools and Applications*, 1–21.

25 Rao, S. and Wang, H. (2021). Robust optical flow estimation via edge preserving filtering. *Signal Processing: Image Communication* 96 (1): 116295–116309.

26 Li R, Tan R. T., Cheong L. F. Robust optical flow in rainy scenes. European Conference on Computer Vision. 2018: 288-304.

6

Advance Deep Learning Approaches for Optical Flow Estimation

6.1 Introduction

Optical flow estimation has long been an important task in computer vision, with applications in fields such as video analysis, autonomous driving, robotics, and augmented reality. Traditionally, variational methods such as the Horn–Schunck and Lucas–Kanade algorithms were used to estimate motion between consecutive image frames. While these methods laid the foundation for motion estimation, they often struggle in complex real-world scenarios, particularly when dealing with large displacements, occlusions, motion blur, or adverse weather conditions.

With the advent of deep learning, optical flow estimation has seen significant advances. By leveraging large datasets and learning from patterns in the data, neural networks can extract hierarchical features that improve the accuracy and robustness of optical flow predictions. This chapter delves into the evolution of deep learning-based optical flow methods, from early architectures like FlowNet to more recent state-of-the-art models such as PWC-Net and RAFT. We also discuss the shift from supervised learning approaches, which rely on labeled datasets, to unsupervised and self-supervised methods that eliminate the need for explicit ground truth optical flow data.

6.2 Convolutional Neural Networks for Optical Flow Estimation

6.2.1 FlowNet: The Pioneer in Deep Learning-based Optical Flow

FlowNet, proposed by [1], was the first deep learning model specifically designed for optical flow estimation. The architecture was inspired by the success of convolutional neural networks (CNNs) in image classification tasks and adapted to handle motion estimation. FlowNet directly predicts the optical flow between two

Robust Optical Flow Estimation in Outdoor Environments, First Edition. Sana Rao and Rao Kashif.

consecutive frames by processing the input images through a series of convolutional layers, followed by a decoder that upsamples the low-resolution feature maps to produce a dense optical flow field.

FlowNet comes in two variants:

- **FlowNetS** (simple) processes two input images as separate channels and then concatenates them before passing through the convolutional layers.
- **FlowNetC** (correlation) computes feature-level correlations between the two input images, which is more suited for motion estimation.

Although FlowNet marked a breakthrough, it struggled with large displacements and fine motion details due to the limitations of its simplistic architecture. Nevertheless, it opened the door for more advanced models that addressed these challenges.

6.2.2 PWC-Net: Pyramid, Warping, and Cost Volume

PWC-Net, introduced by [2], built on the ideas of FlowNet but introduced several innovations to address the limitations in capturing large displacements and handling high-resolution details. The key innovations in PWC-Net include:

1) **Feature Pyramid**: Instead of processing images at full resolution, PWC-Net constructs a pyramid of increasingly lower-resolution images. This allows the network to estimate large motions in coarse levels and refine them as it moves to finer levels.
2) **Warping**: To improve accuracy, PWC-Net warps one image based on the current flow estimate and computes the difference with the second image. This step ensures that only small residual motions need to be estimated at each pyramid level.
3) **Cost Volume**: PWC-Net computes a cost volume at each level, which helps the network understand the spatial relationships between pixels in the two input images. This allows the model to more accurately predict motion, especially in regions with large displacements.

The combination of these techniques led PWC-Net to outperform FlowNet, particularly in complex scenes with fast motion and occlusions.

6.2.3 RAFT: Recurrent All-Paris Field Transforms

RAFT, developed by [3], pushed the boundaries of deep learning-based optical flow estimation even further. Unlike previous models that rely on coarse-to-fine strategies, RAFT operates directly on high-resolution features and employs a recurrent structure to iteratively refine the optical flow estimate. The key components of RAFT include:

- **All-Paris Field Transforms**: RAFT computes a dense cost volume by comparing every pixel in the first image to every pixel in the second image. This all-pairs comparison allows the model to capture both local and global motion patterns.
- **Recurrent Refinement**: Instead of producing a single optical flow estimate, RAFT refines the flow field over multiple iterations using a gated recurrent unit (GRU). This allows the network to focus on difficult regions such as occlusions and fine motion details, improving accuracy with each iteration.
- **Subpixel Precision**: RAFT achieves subpixel accuracy by operating on fine-grained feature representations, allowing it to capture small and subtle motions that other models might miss.

By addressing many of the challenges that plagued earlier models, RAFT has become the state-of-the-art in optical flow estimation, achieving impressive results on benchmarks such as MPI-Sintel and KITTI.

6.3 Unsupervised and Self-supervised Learning Approaches

Supervised learning models like FlowNet and PWC-Net require large amounts of labeled optical flow data for training. However, acquiring such data is expensive and time-consuming, particularly for real-world environments. As a result, unsupervised and self-supervised learning approaches have emerged as promising alternatives.

6.3.1 Unsupervised Learning for Optical Flow

Unsupervised learning methods avoid the need for ground truth optical flow by relying on photometric consistency. The idea is simple: if we warp one image based on the predicted optical flow and compare it to the second image, the two should match. The model is trained to minimize the difference between the warped image and the second image, encouraging it to produce accurate optical flow without the need for explicit supervision.

A key advantage of unsupervised methods is that they can be trained on any video dataset, enabling the model to learn motion patterns from a wide variety of environments. For example, models like UnFlow and DDFlow have been trained on large-scale video datasets, allowing them to generalize to diverse scenes.

6.3.2 Self-supervised Learning and Multi-task Learning

Self-supervised learning takes the unsupervised paradigm further by exploiting auxiliary tasks. For example, in optical flow estimation, a model might be trained to perform both motion estimation and depth prediction. By learning both tasks simultaneously, the model can develop a more comprehensive understanding of the scene, leading to improved performance.

Self-supervised learning has been especially successful in scenarios where labeled data is scarce. The introduction of architectures such as Flow2Stereo and SelfFlow has demonstrated that learning multiple tasks (e.g., depth and motion) can significantly enhance performance without additional ground truth optical flow annotations.

6.4 Conclusion

In this chapter, we have explored the evolution of deep learning approaches for optical flow estimation, from the early successes of FlowNet to the cutting-edge performance of RAFT. The field has made remarkable progress, with models becoming increasingly accurate and capable of handling complex scenarios such as occlusions, motion blur, and adverse weather conditions. As unsupervised and self-supervised methods continue to mature, we can expect further improvements in optical flow estimation, particularly in real-world applications where labeled data is scarce.

Examples with Solutions

Understanding FlowNet Predictions on Synthetic Motion

Problem Statement:
You are given two consecutive synthetic frames of a **20 × 20 pixel grid**, where a small white square moves 3 pixels to the right. Use **FlowNetS** to estimate optical flow and visualize the output flow map.

Solution Steps:

Step 1: Input Pair Definition

We define:

- **Frame 1**: White square at position ($x = 5$–10).
- **Frame 2**: Same square shifted 3 pixels right ($x = 8$–13, $y = 5$–10).

FlowNetS takes these two frames concatenated as a **6-channel input tensor (2 RGB frames)**.

Step 2: Forward Pass through FlowNetS

FlowNetS performs the following:

1) **Convolutional Encoder**: Extracts hierarchical motion features.
2) **Feature Concatenation**: Combines motion cues across spatial scales.
3) **Decoder (Upsampling)**: Produces dense optical flow (u, v).

FlowNetS output for each pixel:

$$F(x, y) = (u, v)$$

where

- u = horizontal motion,
- v = vertical motion.

Step 3: Expected Output

Inside the square:

$u = 3, v = 0$

Background:

$u = 0, v = 0$

Step 4: Interpretation

FlowNetS successfully learns motion correspondence, but minor edge smoothing is visible because FlowNet lacks a cost volume stage (added later in PWC-Net).

PWC-Net Pyramid Matching

Problem Statement:

Demonstrate how PWC-Net improves large-motion estimation using pyramid and warping.

Solution Steps:

Step 1: Multi-scale Image Pyramid

Construct 3 levels:

Level 1 (1/4 resolution)

Level 2 (1/2 resolution)

Level 3 (full resolution)

Step 2: Coarse-Level Flow Estimation

At the coarsest scale, PWC-Net estimates large displacements:

$$u_c = 3, v_c = 0$$

This captures the main shift.

Step 3: Warping

Warp the second image using current flow:

$$I_2'(x) = I_2(x + u_c, y + v_c)$$

Residual motion is now small and easier to refine.

Step 4: Cost Volume
A cost volume $C(x,d)$ is built by correlating patches between warped image and reference:

$$C(x,d) = \sum I_1(x).I_2'(x+d)$$

This captures spatial relationships.

Step 5: Refinement
Fine levels correct residuals, resulting flow is smooth and sharp around object boundaries.

RAFT Iterative Refinement

Problem Statement:
Illustrate how RAFT iteratively refines optical flow on a small grid.

Given:
Two 5×5 patches with displacement $(u, v) = (1, 0)$.

Solution Steps:

1) **All-Pairs Correlation Volume**:
RAFT computes all-pairs similarity between pixel features $f_1(i)$ and $f_2(j)$:

$$C_{ij} = f_1(i).f_2(j)$$

Produces a dense correlation matrix.

2) **Initial Flow Guess**:
Start with $u = v = 0$.

3) **Recurrent Update (GRU)**:
Iteratively updates flow:

$$(u,v)^{t+1} = (u,v)^t + \Delta\,(u,v)$$

Each iteration refines flow based on context.

4) **Subpixel Refinement**:
Continuous representation enables flow precision <1 pixel.

Unsupervised Flow Estimation

Problem Statement:
Train a model to estimate optical flow without ground truth, using photometric consistency.

Solution Steps:

Step 1: Photometric Loss

Warp the second image using predicted flow:

$$I_2'(x) = I_2(x + u, y + v)$$

Loss:

$$L_{photo} = \sum |I_1(x) - I_2'(x)|$$

Model minimizes this to align frames.

Step 2: Smoothness Regularization

Encourage local smoothness:

$$L_{smooth} = \sum |\nabla u| + |\nabla v|$$

Step 3: Combined Loss

$$L = L_{photo} + \lambda L_{smooth}$$

Self-supervised Multi-task Learning

Problem Statement:

Show how combining optical flow + depth estimation helps model learn robust motion cues.

Solution Steps:

1) **Network Outputs**: Flow (u,v) + Depth (D)
2) **Consistency Constraint**:
   Predict scene structure such that:

$$I_2' = Warp(I_1, D, u, v)$$

3) **Loss**:

$$L = L_{photo} + \alpha L_{depth}$$

Interpretation:

Model learns 3D-aware motion estimation, improving performance under occlusions.

References

1 Dosovitskiy, A., Fischer, P., Llg, E. et al. (2015). FlowNet: learning optical flow with convolutional networks. *IEEE International Conference on Computer Vision (ICCV)*, Dec. 7 2015 to Dec. 13 2015 Santiago, Chile. 2758–2766.

2 Sun, D., Yang, X., Liu, M. et al. (2018). PWC-Net: CNNs for optical flow using pyramid, warping, and cost volume. *IEEE Conference on Computer Vision and Pattern Recognition (CVPR)*, June 18 2018 to June 23 2018 Salt Lake City, UT, USA. 8934–8943.

3 Teed, Z. and Deng, J. (2020). RAFT: recurrent all-pairs field transforms for optical flow. *European Conference on Computer Vision (ECCV)*, 23-28 August 2020 Online. 402–419.

7

Optical Flow in Real-Time Systems and Autonomous Vehicles

7.1 Introduction

In recent years, optical flow has become integral to the development of real-time systems, particularly within the domain of autonomous driving. As autonomous vehicles shift from experimental stages to more widespread real-world applications, there is a growing demand for algorithms capable of processing large volumes of visual data in real time. One of the most challenging tasks for autonomous systems is ensuring safe navigation in unpredictable environments. This necessitates accurate motion estimation to detect obstacles, assess relative speeds, and predict future movements. Optical flow is a technique that has been widely adopted in autonomous vehicle systems to address these needs.

As its core, optical flow calculates the movement of objects between consecutive frames of an image sequence. For an autonomous vehicle, this information is critical for interpreting its surroundings. By understanding how the position of various objects changes from frame to frame, the vehicle can make informed decisions about when to stop, steer, or accelerate. This real-time responsiveness is key to achieving advance driver assistance systems (ADAS), which aim to prevent collisions, guide lane changes, and enhance situational awareness.

In this chapter, we will explore how optical flow is applied to real-time systems, with a specific focus on autonomous vehicles. We will cover its role in lane detection, object tracking, and motion prediction, as well as its integration with sensor fusion techniques and deep learning algorithms. We will also discuss the current challenges, such as processing speed, computational efficiency, and adverse weather conditions, and consider future developments that could further advance the field.

Robust Optical Flow Estimation in Outdoor Environments, First Edition. Sana Rao and Rao Kashif.

7.2 Role of Optical Flow in Real-Time Systems

7.2.1 Motion Detection in Dynamic Environments

One of the primary uses of optical flow is motion detection. In dynamic environments, such as urban streets or highways, an autonomous vehicle must continuously monitor moving objects, including pedestrians, cyclists, and other vehicles. Optical flow helps detect the motion of these objects, even when they move in unpredictable ways. For instance, a vehicle using optical flow can detect a pedestrian stepping off the curb or another vehicle suddenly swerving. This allows for real-time responses to potential hazards, improving both safety and efficiency.

The challenge here lies in ensuring the motion detection process is not only accurate but also fast enough to support real-time decision-making. Early optical flow algorithms, such as Horn–Schunck or Lucas–Kanade, while effective, required high computational power and struggled to meet the demands of real-time applications in fast-moving environments. However, more recent algorithms, such as PWC-Net and RAFT, leverage deep learning techniques to significantly improve both speed and accuracy, making them suitable for real-time systems [1].

7.2.2 Lane Detection and Lane-keeping Assistance

For autonomous vehicles, maintaining an accurate understanding of the vehicle's position relative to lane boundaries is crucial. Optical flow is used to enhance lane detection systems by tracking the movement of lane markers in relation to the vehicle's forward motion. Even when the lane markings are faint or partially obstructed by environmental factors such as rain, snow, or heavy traffic, optical flow can estimate lane boundaries by detecting motion patterns from consecutive frames.

Optical flow also assists in lane-keeping assistance systems, which are a part of modern ADAS. These systems ensure that the vehicle remains within its lane by continuously analyzing optical flow data to adjust the steering accordingly. Advanced optical flow models, which combine motion estimation with deep learning, provide highly accurate lane-tracking capabilities even in challenging environments, such as sharp curves or intersections where traditional lane markers may be absent.

7.2.3 Path Planning and Obstacle Avoidance

The ability of an autonomous vehicle to plan its path and avoid obstacles is another critical function enabled by optical flow. In real-time systems, optical flow provides the data needed for dynamic path planning the process by which a vehicle adjusts its route based on changes in its surroundings. For instance, if another vehicle abruptly changes lanes, the optical flow data will capture this motion, allowing the autonomous vehicle to react by altering its own path to avoid a collision.

In scenarios where obstacles are stationary (e.g., construction cones or parked vehicles), optical flow helps the system recognize that the objects are static, allowing for efficient path planning around them. Conversely, for moving obstacles such as other vehicles or pedestrians, the flow vectors generated by optical flow inform the vehicle of their direction and speed, enabling precise motion prediction.

Furthermore, by combining optical flow with stereo vision (using two or more cameras to estimate depth), autonomous vehicles can improve their 3D understanding of the environment. This integration of depth and motion information allows for enhanced obstacle detection, as the vehicle can calculate not only where objects are in space but also how fast they are moving, thus optimizing path planning [2].

7.3 Real-Time Optical Flow in Autonomous Vehicles

Autonomous vehicles must process vast amounts of data in real time to ensure safe and reliable operation. Optical flow plays a significant role in this process, particularly in the perception layer of autonomous driving systems, which involves interpreting sensor data to understand the vehicle's surroundings. Key functionalities supported by optical flow include collision avoidance, pedestrian detection, and real-time environment mapping.

7.3.1 Collision Avoidance and Obstacle Detection

Optical flow provides essential input for collision avoidance systems by detecting motion and predicting the trajectory of nearby objects. When combined with additional sensory data, such as those from LIDAR or RADAR, optical flow algorithms can more accurately assess potential threats and determine whether evasive actions are required.

For instance, if a pedestrian is crossing the street in front of the vehicle, optical flow can quickly identify the movement, estimate the pedestrian's speed and direction, and determine whether the vehicle needs to slow down or stop. This ability to predict object motion is one of the core strengths of optical flow in collision avoidance systems [3].

In addition to pedestrian detection, optical flow helps autonomous vehicles detect obstacles that may not be easily discernible by other sensors. For example, objects that are partially hidden or poorly lit can still be detected by optical flow, as the algorithm is sensitive to even subtle movements. This is particularly useful in urban environments where obstacles, such as cars parked along the roadside or cyclists weaving through traffic, can appear unexpectedly.

7.3.2 Scene Reconstruction and Mapping

In real-time systems, accurate scene reconstruction is essential for autonomous vehicles to navigate effectively. By analyzing the optical flow of consecutive frames, the vehicle can create a detailed map of its surroundings, which includes static objects like buildings and dynamic objects like other vehicles. This map is updated continuously as the vehicle moves, allowing it to track changes in the environment in real time.

One of the challenges of scene reconstruction using optical flow is dealing with occlusions, where one object in the frame obscures another. In these cases, traditional optical flow algorithms may struggle to correctly estimate motion, leading to errors in the reconstructed map. To address this, advanced optical flow models incorporate occlusion detection mechanisms, which identify regions of the image where occlusion occurs and adjust the flow estimation accordingly [4].

7.3.3 Pedestrian and Cyclist Detection

Detecting vulnerable road users, such as pedestrians and cyclists, is one of the most critical tasks for an autonomous vehicle. Optical flow provides valuable information for identifying and tracking the movement of these users in complex urban environments. Real-time detection is essential for the safety of both the vehicle's occupants and other road users.

Recent advancements in pedestrian detection algorithms leverage deep learning models, such as convolutional neural networks (CNNs), which are trained on vast datasets of pedestrian motion patterns. These models, when combined with optical flow data, can detect pedestrians even in crowded or poorly lit environments, ensuring that the vehicle responds appropriately to prevent accidents [3].

7.4 Sensor Fusion and Real-Time Performance Optimization

In autonomous vehicles, sensor fusion refers to the integration of data from multiple sources, such as cameras, LIDAR, and RADAR, to create a more comprehensive understanding of the environment. Optical flow plays a critical role in this process by providing detailed motion data that complements the information from other sensors.

7.4.1 LIDAR and Optical Flow

LIDAR sensors provide highly accurate depth information by emitting laser pulses and measuring the time it takes for them to return. When combined with optical flow, LIDAR can enhance obstacle detection and scene reconstruction by providing precise depth measurements for the objects detected by the optical flow algorithm. This fusion allows the autonomous vehicle to distinguish between static and dynamic objects more accurately and make better-informed decisions about how to navigate around them.

7.4.2 RADAR and Optical Flow

RADAR, which uses radio waves to detect objects, is another sensor commonly used in autonomous vehicles. While RADAR excels at detecting objects in adverse weather conditions, such as heavy rain or fog, it lacks the fine spatial resolution of cameras or LIDAR. By integrating optical flow data with RADAR, autonomous vehicles can improve the resolution of their perception systems, allowing for more accurate detection of small or distant objects that RADAR alone might miss.

7.4.3 Real-Time Performance Optimization

Achieving real-time performance in autonomous vehicles requires optimizing the computational efficiency of optical flow algorithms. One of the main challenges is the trade-off between accuracy and speed. While high-accuracy optical flow models, such as RAFT, offer excellent performance in terms of motion estimation, they can be computationally expensive, making them difficult to implement in real-time systems. To address this issue, researchers have developed pruned neural networks and model compression techniques that reduce the complexity of optical flow models without significantly sacrificing accuracy [1].

Another approach to real-time performance optimization involves the use of hardware accelerators, such as graphics processing units (GPUs) or FPGAs. These specialized processors can perform the intensive calculations required for optical flow at much faster speeds than standard central processing units (CPUs), enabling real-time execution of even the most complex models.

7.5 Challenges and Future Directions

While significant progress has been made in applying optical flow to real-time systems and autonomous vehicles, several challenges remain.

7.5.1 Adverse Weather Conditions

One of the most persistent challenges is ensuring robust optical flow performance in adverse weather conditions. Rain, fog, snow, and low light can all degrade the quality of the images captured by the vehicle's cameras, leading to inaccuracies in the optical flow estimation. To address this, researchers are exploring methods such as image enhancement and noise filtering, which improve the visibility of objects in poor weather conditions [2].

7.5.2 Occlusions and Object Interactions

Handling occlusions remains another major challenge. When one object obscures another, traditional optical flow models often fail to correctly estimate motion, leading to errors in scene reconstruction. Advanced techniques that incorporate occlusion detection are helping to mitigate this problem, but further research is needed to improve the robustness of these methods, particularly in complex, dynamic environments.

7.5.3 Real-Time System Scalability

As autonomous vehicle fleets expand, ensuring the scalability of real-time systems will become increasingly important. Current optical flow algorithms are optimized for individual vehicles, but scaling these systems to handle large fleets of autonomous vehicles operating simultaneously presents new challenges. Cloud-based processing and edge computing are two potential solutions for scaling real-time optical flow systems.

7.6 Conclusion

Optical flow is a key enabler of real-time systems, especially in the context of autonomous vehicles. Its ability to estimate motion, detect obstacles, and predict object trajectories allows for safer, more efficient navigation in dynamic environments. However, challenges such as handling adverse weather conditions, processing occlusions, and optimizing for real-time performance remain. As research in optical flow continues to advance, particularly with the integration of sensor fusion and deep learning techniques, the future of autonomous vehicles looks promising, with further improvements in safety, reliability, and scalability on the horizon.

Examples with Solutions

Time-to-collision (TTC) from optical flow and LIDAR

Problem:
A frontal object has LIDAR-measured distance $D = 20\,\text{m}$. From optical flow, the object's image-space radial expansion suggests relative speed estimate $v_{optical} = \frac{4\text{m}}{\text{s}}$ (positive toward ego vehicle). Combine LIDAR and optical estimate to compute time to collision (TTC) and a fused speed estimate using simple weighted average: weight LIDAR 0.7, optical 0.3. Compute TTC and comment.

Solution:
Given fused speed:

$$v_{fused} = 0.7 \cdot v_{lidar} + 0.3 \cdot v_{optical}$$

But LIDAR gives only distance; we need LIDAR speed estimate. If LIDAR reports relative speed via Doppler or successive scans assume LIDAR speed $v_{lidar} = \frac{3.5\text{m}}{\text{s}}$.

Then:

$$TTC = \frac{distance}{speed} = \frac{20}{3.65} \approx 5.479\ \text{s}$$

Interpretation:
With ~5.5 s to collision, there is moderate time to react. Using fused speed reduces sensitivity to a single sensor's noise. If TTC < 3s, trigger immediate braking; >5s allows planned deceleration.

Forward–Backward Consistency for Occlusion Detection

Problem:
Compute forward–backward (FB) consistency error for a pixel. Forward flow from frame A → B at pixel p is $f = (u_f, v_f) = (2.0, 0.5)$. The backward flow at corresponding location $p' = p + f$ in frame B is b $(u_b, v_b) = (-1.7, -0.4)$. Compute FB error and decide if pixel is occluded using threshold $\tau = 1.0$.

Solution:

Compute $f + b = (2.0 - 1.7, 0.5 - 0.4) = (0.3, 0.1)$.

Norm: $E_{FB} = \sqrt{0.3^2 + 0.1^2} = \sqrt{0.09 + 0.01} = \sqrt{0.10} = 0.3162$.

Compare with threshold $\tau = 1.0$: $0.3162 < 1.0$.

Interpretation:
Low FB error indicates consistent forward/backward estimates; pixel likely not occluded and flow is reliable.

Real-Time Compute Budget: Can RAFT Run at 20 FPS?

Problem:
A deployment platform GPU can do 100 GFLOPS (single-precision) sustained for the optical-flow model. A RAFT-like model requires ~12 GFLOPS per megapixel per inference. The input camera is 640 × 480 (0.3072 MP). Can we run RAFT at 20 frames per second (FPS)? Show calculation.

Solution:
Compute GFLOPS per inference:

$$GFLOPS\,per\,frame = 12\left(\frac{GFLOPS}{MP}\right) \times 0.3072\text{MP} = 3.6864\frac{GFLOPS}{frame}.$$

At 20 FPS:

$$Required\ GFLOPS = 3.6864 \times 20 = 73.728\ GFLOPS.$$

Available GPU throughput = 100 GFLOPS > 73.728 GFLOPS

But consider overheads (pre/post processing, sensor I/O, other tasks) practical margin reduces. Still feasible if model is optimized and pipeline parallelized.

Interpretation:
With 100 GFLOPS GPU, RAFT-like model at this resolution can run ~20 FPS with some headroom; may need model pruning or batching for safety.

Kalman Fusion of RADAR and Optical Flow Velocities

Problem:
A RADAR measured radial velocity $z_R = 8.0$ m/s with variance $R_R = 0.25$ (std 0.5). Optical flow-based velocity estimate $z_O = 7.5$ m/s with variance $R_O = 1.0$ (std 1.0). Using a 1D Kalman update as a sensor fusion (prior state $x_{prior} = 7.8$, prior variance $P_{prior} = 0.5$, perform measurement update by fusing RADAR first then optical flow. Show final fused state and variance.

Solution:
Step A: Update with RADAR ($Z = 8.0$, $R = 0.25$):

Kalman gain:

$$K_R = \frac{P_{prior}}{P_{prior} + R_R} = \frac{0.5}{0.5 + 0.25} = \frac{0.5}{0.75} = 0.6667(0.2) = 7.8 + 0.13333 = 7.93333.$$

State update:

$$x_1 = x_{prior} + K_R(z_R - x_{prior}) = 7.8 + 0.6667(8.0 - 7.8)$$
$$= 7.8 + 0.6667(0.2) = 7.8 + 0.13333 = 7.93333\cdot$$

Covariance update:

$$P_1 = (1 - K_R)P_{prior} = (1 - 0.6667) \cdot 0.5 = 0.3333.0.5 = 0.16667\cdot$$

Step B: Update with Optical flow ($z = 7.5, R = 1.0$):
Kalman gain:

$$K_O = \frac{P_1}{P_1 + R_O} = \frac{0.16667}{0.16667 + 1.0} = \frac{0.16667}{1.16667} = 0.142857\cdot$$

State update:

$$x_2 = x_1 + K_O(z_O - x_1) = 7.93333 + 0.142857(7.5 - 7.93333)$$
$$= 7.93333 + 0.142857(-0.43333) = 7.93333 - 0.0619047 = 7.8714$$

Covariance update:

$$P_2 = (1 - K_O)P_1 = (1 - 0.142857) \cdot 0.16667 = 0.857143.0.16667 = 0.142857\cdot$$

$$\text{Final fused state: } x \approx \frac{7.8714m}{s}\cdot$$

Final variance: P $\approx$ 0.14286 ($std \approx 0.378$)·

Interpretation:
The fused estimate is weighted toward RADAR because RADAR had lower variance; optical flow nudges it slightly downward. Covariance decreased, improving certainty.

References

1 Teed, Z. and Deng, J. (2020). RAFT: recurrent all-pairs field transforms for optical flow. *European Conference on Computer Vision (ECCV)*, 23-28 August 2020 Online. 402–419.

2 Sun, D., Yang, X., Liu, M. et al. (2018). PWC-Net: CNNs for optical flow using pyramid, warping, and cost volume. *IEEE Conference on Computer Vision and Pattern Recognition (CVPR)*, June 18 2018 to June 23 2018 Salt Lake City, UT, USA. 8934–8943.

3 Zhang, C., Chen, Z., Wang, M. et al. (2017). Robust non-local tv-l1 optical flow estimation with occlusion detection. *IEE Transactions on Image Processing* 26 (8): 4055–4067.

4 Dosovitskiy, A., Fischer, P., Llg, E. et al. (2015). FlowNet: learning optical flow with convolutional networks. *IEEE International Conference on Computer Vision (ICCV)*, Dec. 7 2015 to Dec. 13 2015. Santiago, Chile. 2758–2766.

8

Drones and Aerial Surveillance for Optical Flow Estimation in Adverse Outdoor Conditions

8.1 Introduction

8.1.1 The Role of Drones in Modern Surveillance

Drones have become ubiquitous in today's world, where rapid advancements in technology have enabled their use in a range of applications. From monitoring natural disasters to assessing damage in warzones and conducting environmental research, drones are now integral to various sectors. Their ability to fly at varying altitudes, hover in place, and reach otherwise inaccessible areas makes them uniquely suited for aerial surveillance. Furthermore, drones equipped with cameras and sensors can capture high-resolution images and videos, which are invaluable for applications like crowd monitoring, infrastructure inspection, wildlife conservation, and border security.

However, the effectiveness of drone-based surveillance systems largely depends on their ability to process visual data accurately and in real time. One key aspect of this is motion analysis, or detecting and interpreting the movement of objects within a scene. Optical flow estimation is a powerful method used for this purpose. In scenarios where drones are tasked with detecting vehicles in traffic or tracking people in a crowd, precise optical flow estimation allows them to understand motion patterns, making navigation and object tracking more effective.

8.1.2 Challenges in Aerial Surveillance Under Adverse Conditions

While drones offer many advantages, their operation in real-world environments is fraught with challenges, particularly in adverse outdoor conditions. Weather plays a significant role in limiting their performance. Low-light conditions at night or during cloudy days can diminish the drone's ability to detect and track objects. Fog, rain, and snow can obscure the camera's vision, making it difficult

Robust Optical Flow Estimation in Outdoor Environments, First Edition. Sana Rao and Rao Kashif.

to identify objects and measure motion accurately. Moreover, variable altitudes introduce new complexities as the scale of objects in the scene changes. This can make motion detection more difficult, as optical flow algorithms traditionally rely on consistent object sizes between frames.

In addition to environmental factors, the drones themselves are constrained by battery life, computational capacity, and limited onboard storage, all of which limit their ability to process data in real time. The integration of advanced algorithms, such as deep learning-based optical flow techniques, can help overcome some of these challenges by improving motion detection even in low-quality footage.

8.2 Optical Flow for Navigation and Object Tracking

8.2.1 Optical Flow: A Detailed Technical Breakdown

Optical flow refers to the visual motion of objects, edges, and surfaces in a scene caused by the relative movement between the observer (in this case, the drone) and the scene. In computer vision, optical flow is essential for determining how each pixel in a sequence of images moves between frames. It enables the calculation of motion vectors that describe the displacement of objects in the scene. Optical flow has numerous applications, including motion segmentation, object tracking, 3D scene reconstruction, and autonomous navigation.

For drones, optical flow is indispensable. It allows them to interpret the movement of objects within their field of view, enabling dynamic flight adjustments. In particular, optical flow helps stabilize the drone's flight by providing real-time information about how the environment is moving relative to the drone's current position. This is vital in windy conditions or when navigating through cluttered environments, such as urban areas.

In terms of computation, optical flow can be divided into two main approaches:

1) **Dense Optical Flow:** This calculates motion for every pixel in the image, producing a complete motion field that describes how each point in the scene moves between frames. Algorithms such as the Farnebäck method or Horn–Schunck optical flow fall into this category. While dense optical flow provides detailed motion information, it is computationally expensive and may not be feasible for real-time applications on drones with limited processing power.
2) **Sparse Optical Flow:** Instead of calculating motion for every pixel, sparse optical flow focuses on key features in the image, such as corners, edges, or texture-rich areas. A common method for sparse optical flow is the Lucas–Kanade algorithm, which tracks the motion of selected features

between frames. This approach is more efficient for real-time applications and is particularly useful for object tracking in drones.

8.2.2 Optical Flow for Object Tracking: Technical Challenges and Solutions

Object tracking is a core function in aerial surveillance. It involves detecting an object in a scene and following its movement across time. For example, in a traffic monitoring application, the drone's camera would capture a series of frames, and the optical flow algorithm would compute the motion vectors to determine how each car is moving in relation to the drone's position.

However, object tracking using optical flow presents several challenges:

1) **Occlusion:** When an object is temporarily blocked from view by another object (e.g., a car driving behind a building), the optical flow algorithm may lose track of the object, causing inaccuracies in motion estimation.
2) **Scale Variations:** As drones fly at different altitudes, the apparent size of objects in the scene changes. This scale variation can confuse optical flow algorithms, which rely on the assumption that object sizes remain relatively constant between frames.
3) **Camera Shake and Movement:** Drones are susceptible to movement due to wind or sudden flight adjustments. These movements can introduce noise into the optical flow estimation process, causing inaccuracies in object tracking.

To mitigate these challenges, modern optical flow techniques incorporate several improvements:

- **Multi-scale Feature Extraction:** By analyzing the image at multiple scales, algorithms can more effectively handle scale variations. Objects are detected at different resolutions, allowing the drone to maintain accurate motion estimation even as it changes altitude.
- **Motion Compensation:** Advanced techniques compensate for the drone's own movement by isolating the camera's motion from the motion of objects in the scene. This allows for more accurate tracking even when the drone itself is moving unpredictably.
- **Deep Learning Integration:** Deep learning models, especially convolutional neural networks (CNNs), have proven to be highly effective in enhancing object tracking by learning complex motion patterns and compensating for occlusions and scale changes. Deep learning models such as FlowNet and PWC-Net have demonstrated state-of-the-art performance in optical flow estimation, especially under challenging conditions.

8.3 Challenges Regarding Adverse Outdoor Conditions

8.3.1 Environmental Impact on Optical Flow Estimation

Adverse outdoor conditions pose a significant hurdle to accurate optical flow estimation. The environmental conditions that most significantly affect drones during aerial surveillance include:

- **Low-light Conditions:** Optical flow estimation algorithms typically rely on changes in pixel intensity between successive frames to calculate motion. In low-light conditions, such as during nighttime surveillance or cloudy weather, the image's pixel intensities may not change sufficiently between frames, leading to inaccurate or unreliable motion vectors. The absence of sufficient lighting also introduces noise into the image, which can further degrade optical flow accuracy. A simple example can be found in urban traffic surveillance at night, where cars' movements are poorly captured due to insufficient ambient lighting. The optical flow algorithm might incorrectly interpret the movement of headlights or street lamps as motion, confusing object-tracking models.
- **Fog and Rain:** Fog introduces another set of challenges by reducing the contrast and clarity of objects in the scene. The light scattering caused by fog makes it difficult for optical flow algorithms to detect motion accurately, as objects become blurred and lose their distinct edges. In such cases, the motion of fog particles themselves can be detected as noise, obscuring the true motion of objects. Rain and snow further complicate optical flow estimation. The individual droplets or snowflakes introduce small, high-frequency noise that confuses traditional algorithms, making it difficult to estimate motion properly. Moreover, the reflective surfaces caused by wet roads can introduce further artifacts that degrade optical flow computation.
- **Wind and Turbulence:** Strong winds can cause the drone to sway, leading to irregular camera movements. Such external forces impact the drone's stability and make it challenging to differentiate between the drone's movement and object motion. Optical flow models need to account for this additional layer of complexity to prevent erroneous motion estimation.

8.3.2 Occlusion and Noise in Real-World Applications

Real-world environments are rarely as clear-cut as the controlled scenarios often depicted in laboratory settings. In the field, the visual scene is frequently cluttered with objects that partially or fully block the camera's view, known as occlusion. For example, when a drone flies over a crowded city street, pedestrians may block each other from view, or buildings might obscure vehicles that the drone is trying to track.

Traditional optical flow methods often struggle with such occlusions because they rely on consistent visual information across frames. When an object is blocked from view, the algorithm may "lose" the object, causing inaccurate motion predictions.

A related issue is visual noise, which can stem from a variety of sources:

- **Sensor Noise:** Low-quality cameras or adverse weather conditions can introduce random pixel fluctuations, making it harder for optical flow algorithms to detect real motion.
- **Dynamic Backgrounds:** Natural scenes, such as forests or oceans, feature constantly moving backgrounds. In such cases, the challenge is to differentiate between the motion of the background (e.g., trees swaying in the wind) and the motion of objects that are of interest (e.g., a car driving through the forest).

8.4 Improving Optical Flow Estimation for Adverse Conditions

8.4.1 Integration of Deep Learning Models

Recent advancements in deep learning have revolutionized the field of optical flow estimation. CNNs have been particularly successful in learning complex, nonlinear relationships within image sequences. These models can be trained on large datasets of real-world scenarios, enabling them to handle noise, occlusion, and adverse environmental conditions more effectively than traditional methods.

Several deep learning-based approaches have been developed to address the unique challenges of adverse outdoor conditions:

- **FlowNet:** One of the first deep learning architectures designed for optical flow estimation, FlowNet takes two input frames and learns to estimate the flow field directly. The model is trained end-to-end on large datasets and is capable of handling noise and occlusion better than traditional optical flow algorithms.
- **PWC-Net:** A more recent model, PWC-Net, improves upon FlowNet by employing a pyramid, warping, and cost volume approach. The pyramid structure allows the model to process images at multiple scales, improving its ability to estimate motion in the presence of scale variations. The cost volume technique helps the model differentiate between object motion and background noise more effectively.

8.4.2 RADAR and Optical Flow

RADAR, which uses radio waves to detect objects, is another sensor commonly used in autonomous vehicles. While RADAR excels at detecting objects in adverse weather conditions, such as heavy rain or fog, it lacks the fine spatial resolution of cameras or LIDAR. By integrating optical flow data with RADAR, autonomous vehicles can improve the resolution of their perception systems, allowing for more accurate detection of small or distant objects that RADAR alone might miss.

8.4.3 Real-Time Performance Optimization

Achieving real-time performance in autonomous vehicles requires optimizing the computational efficiency of optical flow algorithms. One of the main challenges is the trade-off between accuracy and speed. While high-accuracy optical flow models, such as RAFT, offer excellent performance in terms of motion estimation, they can be computationally expensive, making them difficult to implement in real-time systems. To address this issue, researchers have developed pruned neural networks and model compression techniques that reduce the complexity of optical flow models without significantly sacrificing accuracy [1].

Another approach to real-time performance optimization involves the use of hardware accelerators, such as GPUs or FPGAs. These specialized processors can perform the intensive calculations required for optical flow at much faster speeds than standard CPUs, enabling real-time execution of even the most complex models.

8.5 Challenges and Future Directions

While significant progress has been made in applying optical flow to real-time systems and autonomous vehicles, several challenges remain.

8.5.1 Adverse Weather Conditions

One of the most persistent challenges is ensuring robust optical flow performance in adverse weather conditions. Rain, fog, snow, and low light can all degrade the quality of the images captured by the vehicle's cameras, leading to inaccuracies in the optical flow estimation. To address this, researchers are exploring methods such as image enhancement and noise filtering, which improve the visibility of objects in poor weather conditions [2].

8.5.2 Occlusions and Object Interactions

Handling occlusions remains another major challenge. When one object obscures another, traditional optical flow models often fail to correctly estimate motion, leading to errors in scene reconstruction. Advanced techniques that incorporate occlusion detection are helping to mitigate this problem, but further research is needed to improve the robustness of these methods, particularly in complex, dynamic environments.

8.5.3 Real-Time System Scalability

As autonomous vehicle fleets expand, ensuring the scalability of real-time systems will become increasingly important. Current optical flow algorithms are optimized for individual vehicles, but scaling these systems to handle large fleets of autonomous vehicles operating simultaneously presents new challenges. Cloud-based processing and edge computing are two potential solutions for scaling real-time optical flow systems.

8.6 Conclusion

Optical flow is a key enabler of real-time systems, especially in the context of autonomous vehicles. Its ability to estimate motion, detect obstacles, and predict object trajectories allows for safer, more efficient navigation in dynamic environments. However, challenges such as handling adverse weather conditions, processing occlusions, and optimizing for real-time performance remain. As research in optical flow continues to advance, particularly with the integration of sensor fusion and deep learning techniques, the future of autonomous vehicles looks promising, with further improvements in safety, reliability, and scalability on the horizon.

References

1 Teed, Z. and Deng, J. (2020). RAFT: recurrent all-pairs field transforms for optical flow. *European Conference on Computer Vision (ECCV),* 402–419.
2 Sun, D., Yang, X., Liu, M. et al. (2018). PWC-Net: CNNs for optical flow using pyramid, warping, and cost volume. *IEEE Conference on Computer Vision and Pattern Recognition (CVPR)*, 8934–8943.

9

Conclusions and Future Work

9.1 Conclusions

In this book, we have proposed several novel methods to improve the performance of optical flow estimation, specifically focusing on the preservation of motion boundaries in complex outdoor conditions. The accurate estimation of optical flow is crucial in various applications, such as object tracking, motion segmentation, autonomous vehicles, and video compression. One of the main challenges in these applications is ensuring the robustness and accuracy of motion boundaries in highly dynamic environments. Our work addresses this by introducing multiple robust methods for optical flow estimation, leveraging the nonlocal total variation (NLTV-L^1) framework and its enhancements.

We have emphasized the importance of motion boundaries, as they play a significant role in segmenting moving objects from their backgrounds, which is vital for real-world applications such as autonomous driving, human motion analysis, and surveillance systems. In these environments, challenges like motion blur, low contrast, occlusion, and environmental distortions such as fog and rain can severely degrade the performance of traditional optical flow methods. Our aim throughout this book has been to develop methods that are capable of overcoming these issues by robustly estimating optical flow and preserving critical motion details under adverse outdoor conditions.

At the core of this work is the RNLTV-L^1 method, a robust NLTV-L^1 optical flow estimation method, which we introduced to solve the problems associated with ill-defined edges a blurred motion boundary. This method demonstrates strong performance in preserving motion boundaries, which is a key weakness in many traditional optical flow techniques. By applying the RNLTV-L^1 method to tasks such as motion segmentation, we have shown that motion estimation and segmentation are intrinsically interrelated tasks. Motion boundaries can be effectively used to improve segmentation accuracy, while a more precise segmentation can assist in robust optical flow estimation.

Robust Optical Flow Estimation in Outdoor Environments, First Edition. Sana Rao and Rao Kashif.

We further proposed a robust support weight function to address the over-segmentation problem near motion boundaries. Traditional methods often suffer from over-smoothing in areas with sharp motion discontinuities, leading to inaccurate flow estimations in these regions. The support weight function mitigates this by assigning different weights to neighboring pixels based on their motion and intensity differences. This results in better edge preservation, which is crucial in scenarios involving complex object motion or sudden changes in the scene.

Another significant contribution of this book is the introduction of the NLTV-L^1+WGF (weighted guided filter) method, which builds on the RNLTV-L^1 method by incorporating a weighted guided filter to further enhance the robustness of the optical flow estimation process. This method addresses the limitations of local filtering methods that fail to preserve sharp edges near motion boundaries. The WGF, a widely used edge-preserving filter, allows us to maintain the clarity of motion boundaries while smoothing regions where motion is more uniform. By optimizing the optical flow estimation through the coarse-to-fine process, we ensure that the WGF significantly improves the quality of flow estimates in challenging outdoor scenarios, such as those with rain or fog.

The performance of the proposed methods was evaluated on several benchmark datasets, including Middlebury, MPI-Sintel, and CDnet2014, as well as other challenging outdoor datasets. These datasets encompass a wide variety of outdoor conditions, such as heavy rain, snow, fog, and varying light conditions, providing a comprehensive test for the robustness of our methods. Comparative analysis with state-of-the-art optical flow methods demonstrates that the proposed RNLTV-L^1 and NLTV-L^1+WGF methods consistently outperform traditional techniques, particularly in terms of edge preservation and motion boundary accuracy.

Despite the notable success of the NLTV-L^1+WGF method, there remain certain limitations. The most prominent of these is that the WGF, while effective in edge-preserving operations, does not take into account directional information. This leads us to the next advancement in our work: the combination of the weighted guided filter with a nonlocal steering kernel (WGF-NLSK).

The NLTV-L^1+WGF-NLSK method represents a further refinement of our approach to optical flow estimation. The WGF maintains its edge-preserving properties, but the addition of the NLSK introduces directional awareness to the filter. This enhancement allows the method to better handle edge directions, leading to improved robustness when estimating optical flow in complex scenes where multiple objects move in different directions. The integration of the steering kernel ensures that edges are preserved not just based on pixel intensity difference but also based on the orientation of those edges, which is particularly important in scenes with intricate motion patterns.

The NLTV-L^1+WGF-NLSK method was evaluated on three challenging datasets and compared with existing state-of-the-art optical flow estimation methods. The results showed that our proposed method was superior in terms of accuracy, particularly in preserving motion boundaries in complex environments. This demonstrates that the introduction of the NLSK was instrumental in achieving the desired robustness, making it a promising solution for real-time applications such as autonomous navigation systems, where preserving motion boundaries is critical.

Overall, this book has successfully demonstrated the potential of combining nonlocal filtering techniques, total variation regularization, and advanced edge-preserving filters to improve the robustness of optical flow estimation under adverse outdoor conditions. Through extensive experiments, we have shown that the proposed methods consistently outperform existing techniques, especially in challenging real-world scenarios where environmental factors such as rain, fog, and snow significantly impact optical flow accuracy.

9.2 Future Work

While we have made significant progress in advancing the state-of-the-art in optical flow estimation, there are several avenues for future research and development that can build on the work presented in this book. The following are some key areas where further improvements can be made:

1) **Development of Larger, More Diverse Datasets:** One of the primary limitations in evaluating optical flow estimation methods, especially for adverse outdoor conditions, is the lack of large, high-quality datasets with ground-truth flow values. Many existing datasets, such as Middlebury and MPI-Sintel, provide valuable benchmarks but are limited in the range of environmental conditions they cover. Future work should focus on developing new datasets that encompass a broader spectrum of outdoor conditions, including fog, haze, heavy rain, and varying light conditions. These datasets should also provide ground-truth optical flow values for quantitative evaluation, enabling more rigorous comparisons between different methods.
2) **Real-Time Optical Flow Estimation:** As optical flow estimation is increasingly applied in real-time systems, such as autonomous vehicles and drones, there is a growing need to improve the computational efficiency of these methods. While the methods presented in this book demonstrate robustness and accuracy, their computational complexity can be a barrier to real-time application. One potential solution is to implement these methods on graphics processing units (GPUs) to take advantage of parallel processing and significantly speed up computation. Another approach is to explore the use

of lightweight neural networks that can approximate the behavior of more complex methods while maintaining acceptable accuracy.

3) **Improving Robustness in Extreme Environmental Conditions:** Despite the improvements introduced by the NLTV-L^1+WGF-NLSK method, there are still extreme environmental conditions, such as heavy snowfall, sandstorms, and dense fog, where optical flow estimation remains a challenging task. Future work could explore the integration of deep learning techniques with traditional variational methods to improve robustness under such conditions. For instance, convolutional neural networks (CNNs) could be trained to recognize specific environmental artifacts and compensate for them during the flow estimation process.
4) **Advanced Driver Assistance Systems (ADAS) and Autonomous Driving:** The development of robust optical flow estimation methods is closely tied to the advancement of autonomous driving systems. However, current optical flow methods still struggle with environmental challenges like rain, fog, and sudden lighting changes, which can degrade the accuracy of motion estimation. Future research should focus on optimizing these methods for deployment in ADAS, where reliability and real-time performance are critical. Additionally, more work is needed to ensure that these methods can handle rare but dangerous situations, such as black ice on the road or sudden weather changes.
5) **Multi-model Optical Flow Estimation:** Another promising direction for future research is the integration of multi-model sensor data to improve optical flow estimation. In addition to visual data, information from depth sensors, infrared cameras, and radar could be used to provide a more comprehensive understanding of the scene. For example, depth data from LIDAR could be combined with optical flow estimates to improve accuracy in scenarios where traditional vision-based methods struggle, such as in low-light environments or when there is significant occlusion.
6) **Application to New Domains:** While much of the focus in this book has been on autonomous driving and motion segmentation, optical flow has numerous other potential applications. For example, it could be applied to medical imaging, where the estimation of motion between frames in an MRI or ultrasound scan could help detect anomalies in tissue movement. Similarly, sports analytics could benefit from more accurate motion estimation to track athletes' movements and improve performance analysis. Extending the methods presented in this book to these new domains could open up exciting new avenues for research and application.
7) **Hybrid Approaches Combining Classical and Deep Learning Methods:** Another direction for future research is the development of hybrid approaches that combine the best aspects of classical variational methods with the flexibility and learning capacity of deep learning models. For example, a deep

network could be trained to estimate optical flow but be guided by classical priors, such as total variation or nonlocal filtering, to ensure robustness in challenging scenarios. These hybrid models could strike a balance between the interpretability and stability of traditional methods and the high performance of modern machine learning techniques.

In conclusion, while we have made substantial progress in advancing the field of optical flow estimation, there remains a wealth of potential for future exploration and development. By continuing to push the boundaries of what is possible with optical flow estimation, we can unlock new possibilities for computer vision applications across a wide range of domains.

Index

Robust Optical Flow Estimation in Outdoor Environments, First Edition. Sana Rao and Rao Kashif.